Lecture Notes in Physics

Edited by H. Araki, Kyoto, J. Ehlers, München, K. Hepp, Zürich
R. Kippenhahn, München, H. A. Weidenmüller, Heidelberg
and J. Zittartz, Köln

Managing Editor: W. Beiglböck, Heidelberg

225

Berndt Müller

The Physics of the Quark-Gluon Plasma

Springer-Verlag
Berlin Heidelberg New York Tokyo

Author

Berndt Müller
Institut für Theoretische Physik der Johann-Wolfgang-Goethe-Universität
Postfach 11 19 32, D-6000 Frankfurt/Main, F.R.G.

ISBN 3-540-15211-3 Springer-Verlag Berlin Heidelberg New York Tokyo
ISBN 0-387-15211-3 Springer-Verlag New York Heidelberg Berlin Tokyo

Printed in Germany

Printing and binding: Beltz Offsetdruck, Hemsbach/Bergstr.
2153/3140-543210

FOREWORD

These notes have grown out of a set of lectures given at the Université de Liège in Spring 1983 for the 3ème Cycle FNRS en Physique Nucléaire. When several members of the theory group at Liège suggested that it would be nice to have the lecture notes available in printed form, I thought this would be an easy thing to do. As the writing-up progressed, however, it became increasingly clear that even for some of the basic questions about the physics of the quark-gluon plasma no definite answer was available in the literature. Some of these questions are: Of what type is the phase transition from hadronic to quark matter? Can there be overheating or undercooling resulting in large fluctuations? How does a fireball consisting of quark-gluon plasma finally break up: by bulk expansion, by surface radiation, or by successive fissioning? What is the best experimental signature for the existence of the quark-gluon plasma, with quantitative predictions including background processes?

In several cases simple calculations had to be done in order to be able to give even a crude estimate of the importance of specific processes or effects. Such considerations concern, e.g.:

- pion radiation from the surface of the plasma (in Chap.4),
- the possible lifetime of a supercooled quark-gluon plasma phase in the case of a first-order phase transition (in Chap.7),
- the analysis of the momentum-dependence of the running coupling constant $\alpha_s(q,T)$ at finite temperature (in Chap.8),
- the influence of finite bag size on the apparent number of degrees of freedom of the quark-gluon plasma (in Chap.9).

The specialist will certainly find that some (possibly his or her own) contributions to the published literature have not received their due attention and presentation. For such deficiencies as well as all other misrepresentations and omissions, in particular from the list of references, the author takes all the blame. Nevertheless, he hopes that these lecture notes may be useful as an introduction to the theory of the quark-gluon plasma, which contains a lot of beautiful and interesting physics, and - perhaps - as a motivation for the reader to contribute to the solution of the many open questions.

I wish to thank all members of the nuclear theory group at Liège, in particular M.Bawin, J.Cugnon, J.Humblet, C.Mahaux, F.Stancu, and also H.Caprasse, for their kind invitation to give these lectures, their interest, and their stimulating questions. I also would like to thank those who have aroused my interest in this field and from whose knowledge I have benefitted immensely, especially J.Rafelski and his students (T.Elze, P.Koch, A.Schnabel, and J.Staadt), M.Gyulassy, U.Heinz, L.McLerran, and J.Zimányi.

Frankfurt, December 1983.

Preface to the Augmented Version

When these lecture notes were originally prepared, the perspective for nuclear collision experiments at very high energy was still uncertain. In the meantime, experimental efforts to search for the quark-gluon plasma are well under way. Experiments with light and medium-heavy nuclei (^{16}O, ^{32}S, ^{40}Ca ?) will be running at CERN and Brookhaven National Laboratory in the years 1986-88. Plans for a high-energy heavy-ion collider to be constructed in the United States are taking shape.

When Springer-Verlag kindly agreed to publish the manuscript, several additions had to be made in order to include more recent developments. The main modification concerns a new chapter on relativistic hydrodynamics of the quark-gluon plasma and its much studied scaling solution. In view of the progress in the planning of experiments, extended amendments have also been made in the chapter on experimental signatures of the quark-gluon plasma. New results have been included in the chapter on Monte-Carlo simulations of the lattice gauge theory. The list of references has been brought up to date. I hope that the publication of the lectures will help to attract further interest to this emerging field of research at the juncture of nuclear and high-energy physics.

Frankfurt, January 1985.

CONTENTS

LECTURE 1

We start with a brief review ("Schnellkurs") of the basic ideas of QCD, the gauge theory of strong interactions among elementary particles carrying colour. We show how QCD is modelled after quantum electrodynamics (QED) and pin-point the differences originating in the non-linearities caused by the noncommutativity of the colour gauge group SU(3). After spelling out the fundamental equations we present the rules for Feynman diagrams in QCD. Applying these rules to the polarization function of the gluon propagator we find the expression for the "running" coupling constant $\alpha_s(q^2)$ exhibiting asymptotic freedom and a strong increase at small momenta.

We then proceed to the MIT-bag model which provides an intuitive framework for colour confinement and hadronic structure . We discuss the boundary conditions for quarks and gluons and the meaning of the bag constant. The bag model allows for a reasonable fit of the low energy spectrum of hadrons with the exception of the pions. Violation of the partial conservation of axial current (PCAC) in the MIT-bag model leads to modifications, the chiral bag models. In these, the pion field is treated as an additional elementary field, with a surface coupling to quarks and antiquarks.

1. QUANTUM CHROMODYNAMICS

Particle physicists believe today that quantum chromodynamics (QCD) is the fundamental theory of strong interactions. What is QCD? From high-energy scattering experiments of leptons (electrons, muons or neutrinos) on hadrons, i.e. strongly interacting particles, we know that there must be almost point-like constituents inside the hadrons. These constituents are commonly called partons, and are identified with the quarks that were first postulated in the context of the classification of hadrons according to the representations of SU(3). From the analysis of lepton-hadron scattering it is known that quarks have spin $\frac{1}{2}$, i.e. they are Dirac particles, and that they carry non-integer electric charge.

Today we know that there are at least five different "flavours" of quarks, called u(p), d(own), s(trange), c(harm) and b(ottom), respectively, and there are strong theoretical reasons to suspect the existence of one more flavour, called t(op). The quarks carry electric charge, which takes the value +2/3 for u,c,(t) and the value -1/3 for d,s,b in units of the proton charge. The quarks of various flavour differ widely by their mass. Because the quarks are confined to the inside of hadrons - at least under normal circumstances - their mass cannot be measured directly. Most of the mass of the hadron itself does not originate in the intrinsic mass of the quark constituents but resides in the kinetic energy of the confined quarks and in the field that binds the quarks together, the so-called glue field. By the methods of current algebra one can extract something like bare quark masses (called masses of "current quarks"), for which we list typical values [LP79] :

$$\begin{aligned} &m_u \simeq 5\text{ MeV}, \quad &&m_d \simeq 10\text{ MeV}, \quad &&m_s \simeq 150\text{ MeV}, \\ &m_c \simeq 1500\text{ MeV}, \quad &&m_b \simeq 5000\text{ MeV}, \quad &&m_t \simeq 40\text{ GeV}. \end{aligned} \tag{1.1}$$

The quarks must carry one further internal quantum number which is called *colour*. This can be deduced from the existence of particles like the Δ^{++} = (uuu), the Δ^- = (ddd) and Ω^- = (sss) which contain three identical quarks with parallel spin in a s-wave. The Pauli principle requires that the quarks differ from each other by an additional quantum number. An independent argument relies on the ratio of the production of hadrons from e^+e^- collisions as compared with the production of muon pairs. If the production of hadrons proceeds through creation of a quark-antiquark pair which subsequently fragments into hadrons, the ratio of the cross-sections can be related to the sum over the square of the charges of all sorts of quarks that can be created. Counting the different colours separately gives a factor N_c for the number of colours:

$$R = \sigma(e^+e^-\to\text{hadrons})/\sigma(e^+e^-\to\mu^+\mu^-) = N_c \sum_i e_i^2 = (11/9)N_c \,. \tag{1.2}$$

for i=u,d,s,c,b. The experiment yields $N_c = 3$ without doubt [PDG82].

We believe today that the quantum number of colour resembles very much in its properties that of electric charge, so much so that one also speaks of "colour charge". Like electric charge, colour charge is exactly conserved, and it acts as the source of a force field which is of long range unless screened. The difference is that there are three different colour degrees of freedom, say red, green and blue, whereas electric charge is a one-dimensional, and therefore additive, quantity. The theory of colour forces, i.e. quantum chromodynamics, is derived from the principle of gauge invariance against arbitrary rotations in colour space. Since the complex rotations of a three-dimensional vector are described by unitary 3×3-matrices $U=(u_{ik})$ of unit determinant, the symmetry group of the gauge transformations is SU(3).

The precise form of the QCD Lagrangian is obtained by considering invariance under local colour rotations, i.e. by allowing the gauge matrix U to change from one point in space to another. To see how this principle of local gauge invariance works, let us go back to quantum electrodynamics (QED). In QED the gauge transformations correspond to changes in the phase of the wavefunction: $\psi \to e^{i\alpha}\psi$. If the phase α varies from point to point, i.e. if $\alpha=\alpha(x)$, the derivative of a wavefunction changes by a non-trivial term:

$$\partial_\mu\psi \;\to\; \partial_\mu(e^{i\alpha(x)}\psi) \;=\; e^{i\alpha}(\partial_\mu\psi) + e^{i\alpha}(i\partial_\mu\alpha)\psi \,. \tag{1.3}$$

The unwanted second term is cancelled by the gauge change of the electromagnetic potential A_μ, if the potential is added to the derivative operator in minimal coupling $(\partial_\mu+ieA_\mu)$. With

$$A_\mu(x) \;\to\; A_\mu(x) - e^{-1}(\partial_\mu\alpha) \tag{1.4}$$

one finds that the minimally coupled derivative remains invariant:

$$(\partial_\mu+ieA_\mu)e^{i\alpha(x)}\psi \;=\; e^{i\alpha(x)}(\partial_\mu+ieA_\mu)\psi \,. \tag{1.5}$$

The electromagnetic field strength tensor $F_{\mu\nu} = \partial_\mu A_\nu - \partial_\nu A_\mu$ also being invariant against the gauge transformations (4), the Lagrangian of QED is:

$$L_{QED} \;=\; i\bar{\psi}\gamma^\mu(\partial_\mu+ieA_\mu)\psi - m\bar{\psi}\psi - \tfrac{1}{4}F^{\mu\nu}F_{\mu\nu} \,. \tag{1.6}$$

The form of this Lagrangian, and therefore the form of the electromagnetic interaction, is uniquely determined by the requirement of local gauge invariance.

Because there are three colours, the wavefunction of a quark has three components in colour space: $\Psi = (\psi_r,\psi_g,\psi_b)$, where the indices r, g, b stand for 'red', 'green', and 'blue'. As we already said, a colour gauge transformation is described by a unitary 3×3-matrix U with det(U)=1, which rotates the colour components of the wavefunction: $\Psi \to U\Psi$. Now any unitary matrix can be written as the imaginary exponential of a hermitian matrix: $U = \exp(iL)$. We have $U^*U=1$ and $L^*=L$, and $\det(U)=1$ means $\mathrm{tr}(L)=0$. All traceless hermitian 3×3-matrices can be expressed as linear combination of the eight λ-matrices of Gell-Mann [GM62]:

$$L = \tfrac{1}{2}\,\Sigma_{a=1}^{8}\,\Theta_a\lambda_a . \tag{1.7}$$

We do not need the explicit form of the λ_a, only the commutation relations

$$[\lambda_a,\lambda_b]_- = 2if_{abc}\lambda_c , \qquad [\lambda_a,\lambda_b]_+ = (4/3)\delta_{ab} + 2d_{abc}\lambda_c . \tag{1.8}$$

where the f_{abc} and d_{abc} are the antisymmetric and symmetric structure constants of the Lie group SU(3), respectively, and summation over c is implied.

We now make the colour rotation U space-dependent by allowing the eight real parameters Θ_a to change from point to point: $U(x) = \exp(\tfrac{1}{2}\Theta_a(x)\lambda_a)$ with summation over 'a' implicitly understood. The partial derivative then acquires an unwanted additional term:

$$\partial_\mu(U(x)\Psi) = U\partial_\mu\Psi + (\partial_\mu U)\Psi = U\,[\partial_\mu\Psi + U^*(\partial_\mu U)\Psi] . \tag{1.9}$$

Instead of the term $(i\partial_\mu\alpha)$ in eq.(3) we have now a matrix term $U^*(\partial_\mu U)$. In order to cancel this contribution we have to introduce a colour potential $\hat{A}_\mu$ which is a 3×3-matrix, indicated by the 'hat' symbol. As the potential must be hermitian, we can represent it as linear combination of the Gell-Mann matrices with eight real potential functions $A_\mu^a(x)$:

$$\hat{A}_\mu(x) = \tfrac{1}{2}\,\Sigma_{a=1}^{8}\,A_\mu^a(x)\lambda_a . \tag{1.10}$$

The field $\hat{A}_\mu(x)$ is called a Yang-Mills field [YM54,Ut56,GG61]. If the potential changes under a local colour rotation according to

$$\hat{A}_\mu \rightarrow U^* \hat{A}_\mu U - ig^{-1} U^* (\partial_\mu U) ,$$

(1.11)

the minimally coupled derivative $(\partial_\mu - ig\hat{A}_\mu)$ remains invariant in form under the gauge transformation:

$$(\partial_\mu - ig\hat{A}_\mu) U\Psi = U(\partial_\mu - ig\hat{A}_\mu)\Psi .$$

(1.12)

That one needs eight colour potentials instead of one is not really a surprise. The three-component quark wavefunction forms a triplet of the colour group SU(3), while the wavefunction of an antiquark forms an antitriplet. From the product of a triplet and an antitriplet one can form a SU(3) singlet or an octet: $3 \times 3^* = 8 + 1$. Since the colour potential must have the same quantum numbers as a quark-antiquark pair and cannot be a colourless singlet, it must be described by an octet of fields.

One can show that the eight-component field strength tensor

$$F^a_{\mu\nu} = \partial_\mu A^a_\nu - \partial_\nu A^a_\mu + g f_{abc} A^b_\mu A^c_\nu$$

(1.13)

remains form-invariant under a local colour gauge transformation, i.e. $F_{\mu\nu} \rightarrow U^* F_{\mu\nu} U$, where the matrix F is defined in the same way from its eight colour components as the matrix $\hat{A}$. The complete Lagrangian of quantum chromodynamics is then [FG72,We73,GW73]

$$L_{QCD} = i\bar{\Psi}\gamma^\mu(\partial_\mu - ig\hat{A}_\mu)\Psi - m\bar{\Psi}\Psi - \tfrac{1}{4}F^a_{\mu\nu}F^{\mu\nu}_a .$$

(1.14)

The similarity to the Lagrangian of QED, eq.(6) is conspicuous. All the difference lies in the nonlinear term entering into the definition of $F^a_{\mu\nu}$, eq.(13), which is quadratic in the colour potentials A^a_μ. Since the antisymmetric structure coefficients f_{abc} vanish when two indices are equal, the nonlinear terms disappear if just a single colour component of A^a_μ is different from zero. It follows directly from this observation that all solutions of classical electrodynamics, where there is only one potential to start with, are also solutions of classical chromodynamics. There are, of course, further solutions if the nonlinear terms contribute.

In the quantum theory one cannot make certain components of the colour potential vanish at will, and therefore the nonlinearities always show up. The solutions of QCD are radically different from those of QED, in particular, free colour charges do not exist. This property, called confinement of colour, means that particles carrying colour charge always come in combinations which form total colour singlets, i.e. all long-range colour forces are completely screened. The proof of these assertions comes from numerical calculations which will be discussed in some detail in ch.6.

The analogue of the Maxwell equations may be derived for the colour field from the Lagrangian (14):

$$\partial_\nu F_a^{\mu\nu} = g j_a^\mu - g f_{abc} A_\nu^b F_c^{\mu\nu} , \tag{1.15}$$

where the colour current of the quarks is obtained from L_{QCD}:

$$j_\mu^a = \tfrac{1}{2}\bar{\Psi}\gamma^\mu \lambda_a \Psi . \tag{1.16}$$

The nonlinear term appearing on the right-hand side of eq.(15) tells that the colour field acts partially as its own source. In other words, the quanta of the colour field, i.e. the *gluons*, carry colour charge themselves. This is the origin of the differences between QCD and QED. From eq.(15) one can see, that j_μ^a does not obey a continuity equation, which means that the colour charge of the quarks alone is not conserved. This is not surprising since quarks can emit or absorb gluons which carry away colour. Only if we add the colour charge residing in the gluon field, represented by the second term on the right-hand side of eq.(15), a conserved colour current is obtained. We finally note that one can derive an energy-momentum tensor

$$T_\mu{}^\nu = -L_{QCD}\delta_\mu{}^\nu - F^a_{\mu\lambda}F^{\nu\lambda}_a \tag{1.17}$$

for the gluon field. Its divergence yields the analogue of the Lorentz force, now the force acting on a particle moving in a colour field:

$$\partial_\nu T^{\mu\nu} = gF^{\mu\nu}_a j^a_\nu . \tag{1.18}$$

The energy density has the familiar form $T_{00} = \frac{1}{2}(E^a)^2+\frac{1}{2}(B^a)^2$, where E^a and B^a are the electric and magnetic components of the colour field strength tensor.

The perturbative quantization proceeds in QCD in much the same way as known from QED, because QCD can be shown to be renormalizable [tH71]. The quadratic terms in the Lagrangian L_{QCD} define free quark and gluon fields which are described by propagators which have the same form as those for electrons and photons in QED. One small difference arises because of the coupling of the gluon field to itself. To ensure gauge invariance of the quantum theory one has to introduce fictitious particles called (Faddeev-Popov) ghosts [FP67], which carry colour but behave like fermions although they propagate like spin-zero particles. These particles cancel the contributions from the unphysical degrees of freedom of the colour gauge field. The terms of third and fourth order in L_{QCD} give rise to interaction vertices among the free propagators of quarks, gluons and ghosts. There is a quark-gluon vertex, three-gluon and four-gluon vertices, and a gluon-ghost vertex. Propagators and vertices can be combined to Feynman diagrams in all possible ways, with the exceptional rule that ghosts, being fictitious particles, can only occur in intermediate states. The propagators and vertices of QCD in the Coulomb gauge are shown graphically and analytically in Fig.1.

The free propagators of the gluon, which is proportional to $1/q^2$, suggests that the colour force falls off like $1/r$. The true gluon propagator, how-

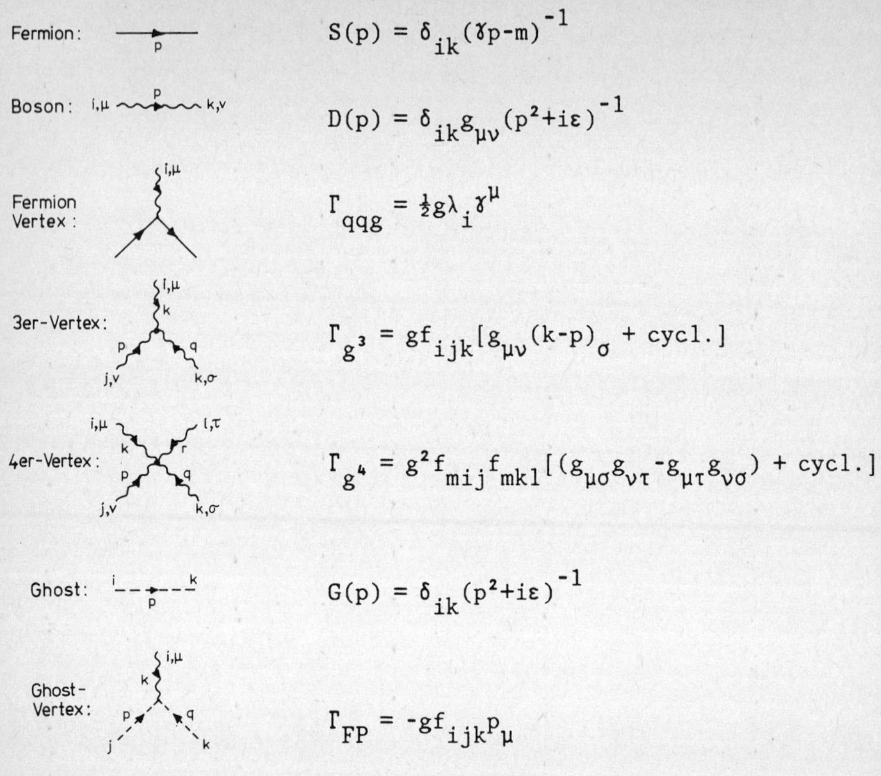

Fig.1: Propagators and vertices of quantum chromodynamics. Ghost lines are not allowed to appear as external lines in Feynman diagrams.

ever, is modified by vacuum polarization and shows a completely different behaviour. In order to see how this comes about, one has to evaluate the loop diagrams corresponding to the virtual creation of a pair of coloured particles from the vacuum:

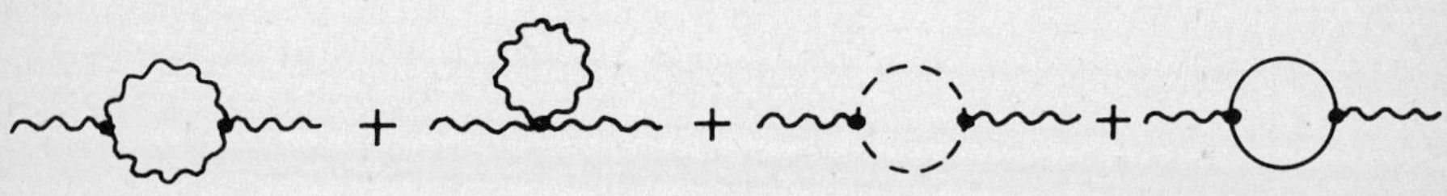

The effect of these diagrams can be expressed in terms of a polarization function

$$\Pi(q^2) = -(33-2N_F)/48\pi^2 \; g^2q^2\ln(-q^2/\mu^2) \; , \tag{1.19}$$

where μ is a reference point introduced by renormalization and N_F counts the number of quark flavours with mass below $|q^2|^{\frac{1}{2}}$. The sign of Π is opposite to that of the polarization function in QED due to the presence of the second diagram shown above, involving the gluon loop. The higher-order diagrams, in which the gluon interacts consecutively once, twice, three times with the vacuum polarization, and so on, can be summed into a geometric series for the full propagator:

$$\sim\!\sim + \sim\!(\Pi)\!\sim + \sim\!(\Pi)\!\sim\!(\Pi)\!\sim + \cdots$$

$$\begin{aligned} D(q^2) &= D_0(q^2) + i^2D_0(q^2)\Pi(q^2)D_0(q^2) + i^4D_0\Pi D_0\Pi D_0 + \ldots \\ &= D_0(q^2)[1-\Pi(q^2)D_0(q^2)]^{-1} \\ &= 1/q^2 \; [1 + (33-2N_F)\alpha_s/12\pi \; \ln(-q^2/\mu^2)]^{-1} \; . \end{aligned} \tag{1.20}$$

After renormalization the factor in brackets acts as a momentum-dependent modification of the strong coupling constant. When it is combined with α_s one obtains the "running" coupling constant

$$\alpha_s(q^2) = 4\pi[(11-2N_F/3)\ln(-q^2/\Lambda^2)]^{-1} \; , \tag{1.21}$$

where Λ is a dimensional parameter introduced in the renormalization process. α_s has vanished altogether from the expression (21), an effect known as "dimensional transmutation" that is commonly found in massless quantum field theories [CW73].

The determination of Λ from experimental data is a somewhat tricky business [St81]. Besides other subtleties this has to do with the different behaviour of (21) for space-like and time-like momenta. For space-like momenta, $q^2<0$, the logarithm in eq. (21) is real, and the measured coupling constant can be identified with $\alpha_s(q^2)$. For time-like momenta, $q^2>0$, on the other hand, the logarithm becomes complex, $\ln(q^2/\Lambda^2)-i\pi$, and the measured value must be identified with $\mathrm{Re}(\alpha_s)$. The occurrence of an imaginary part in α_s corresponds to the possibility of a time-like gluon splitting into two gluons or into two light quarks. Analyses of experiments involving space-like momenta (in particular charmonium level fits) indicate a value for Λ in the range of 300-500 MeV/c [VR80,Mo83]. The dependence of $\mathrm{Re}(\alpha_s)$ for positive and negative q^2 is shown in Fig. 2.

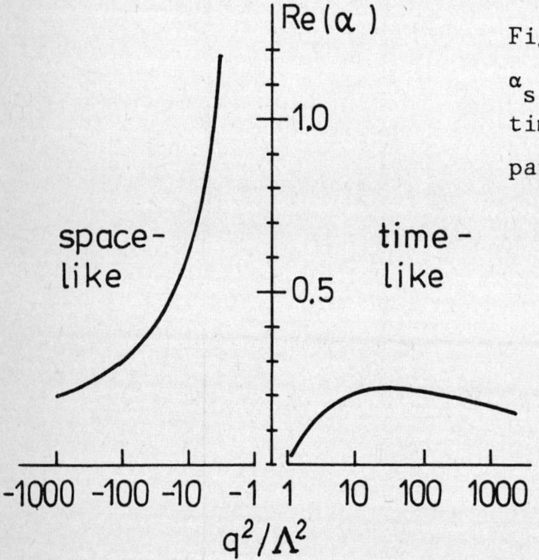

Fig.2: Running coupling constant $\alpha_s(q^2)$ as function of q^2/Λ^2. For time-like momenta only the real part is shown.

At $q^2=-\Lambda^2$ the running coupling constant has a pole. That this pole occurs at a finite value of q^2 is an artefact of the approximation (19) involving only the simplest loop diagram. More sophisticated approximate solutions of the Schwinger-Dyson equations for the gluon propagator [BBZ81] indicate that the pole should really be at $q^2=0$, and that $\alpha_s(q^2)$ should behave as $1/q^2$ in the limit $q^2\to 0$. Converted into coordinate space this would mean that $\alpha_s(r)$ grows like r^2 for large distances, corresponding to a linearly rising potential (see Fig. 3). This implies permanent confinement of quarks and gluons, because an infinite amount of energy would be needed to separate two colour charges.

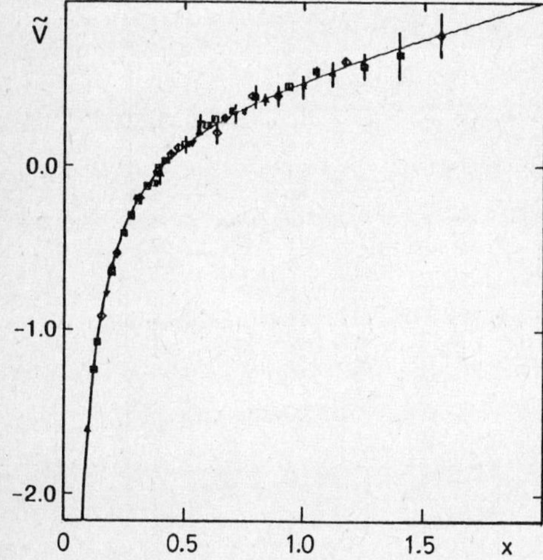

Fig.3: The quark-antiquark potential as obtained from Monte-Carlo calculations on a lattice. Dynamical quark pairs are not included (from [St84a]).

On the other hand, the logarithm in eq. (21) causes a gradual decrease of the coupling strength between colour charges at large momenta or small distances. This is the celebrated property of *asymptotic freedom* which makes QCD the prime candidate for a theory of strong interaction, because it describes the approximate scaling of cross-sections at very high energy.

As we have seen, the strength of the interaction among quarks and other quarks or antiquarks grows steadily with increasing separation. Numerical calculations like those discussed in ch.6 show that the potential grows linearly with distance beyond 1 fm. I.e. in QCD there is a constant long-range force between the quarks, in contrast to QED where the force between charged particles decreases like the square of the distance. This leads, ultimately, to permanent confinement of quarks and, more generally, to confinement of all coloured particles. All free states in QCD correspond to colour singlets. (We shall see later, in ch.9, how this condition can be implemented in a statistical calculation.)

As QCD is a much too complicated theory to allow for simple analytical solutions, models have been developed that allow to incorporate confinement into hadronic structure calculations in a schematic manner. For our purpose the most attractive of these models are the so-called "bag" models which are based on the concept of quarks confined to a given hadronic volume V. Following Bogolioubov [Bo67] confinement is forced upon the quarks by the assumption that their (effective) mass, which is small inside the bag volume, becomes very large outside the bag. Neglecting short-range interactions among the quarks, the quark wavefunctions then must obey the Dirac equation

$$i\gamma^{\mu}\partial_{\mu}\psi - M\psi + (M-m)\theta_{V}\psi \;=\; 0 \;, \tag{2.1}$$

where $\theta_V=1$ inside the bag and $\theta_V=0$ outside. In the limit $M\to\infty$ it is possible to show that the wavefunction vanishes outside the bag volume and satisfies a linear boundary condition at the bag surface [Ch74]:

$$i\gamma^\mu n_\mu \psi|_S = -i(\vec{\gamma}\cdot\vec{n})\psi|_S = \psi|_S \ . \tag{2.2}$$

Because $\gamma^{\mu+} = \gamma^0\gamma^\mu\gamma^0$, the adjoint equation reads $-i\bar{\psi}\gamma^\mu n_\mu|_S = \bar{\psi}|_S$. It is now easily seen that the linear boundary condition ensures that the normal flow of quark current through the bag surface vanishes: $n_\mu j^\mu = n_\mu(\bar{\psi}\gamma^\mu\psi) = 0$. The boundary condition (2) thus guarantees confinement of quarks.

The free Dirac equation $i\gamma^\mu\partial_\mu\psi - m\psi = 0$ with the condition (2) is easily solved for a bag of spherical shape. For m=0 and bag radius R one finds the solutions with energy ω

$$\psi_\kappa^\mu = N \begin{pmatrix} j_\ell(\omega r)\chi_\kappa^\mu \\ \\ iS_\kappa j_{\bar{\ell}}(\omega r)\chi_{-\kappa}^\mu \end{pmatrix} . \tag{2.3}$$

Here $\kappa=\pm(j+\frac{1}{2})$ with j,μ being the angular momentum of the quark, $S_\kappa=\kappa/|\kappa|$, and $\ell=\kappa$ for $\kappa>0$, $\ell=|\kappa|+1$ for $\kappa<0$, $\bar{\ell}=\ell-S_\kappa$. The dimensionless eigenvalue $x=\omega R$ is determined from

$$j_\ell(x) = -S_\kappa j_{\bar{\ell}}(x) . \tag{2.4}$$

For s-states ($\kappa=-1$) we have the equation $j_0(x)=j_1(x)$ with the eigenvalues $x = \omega R = 2.04, 5.40, \ldots$ Perhaps the most spectacular success of this crudest form of the bag model is its prediction of the mass of the "Roper" resonance of the nucleon with mass M_R = 1440 MeV. The assumption is that the nucleon corresponds to a configuration with three quarks in the lowest s-state, while one of the quarks is excited to the second s-state in the Roper resonance. We therefore have

$$M_R = (2\omega_1+\omega_2)/(3\omega_1)\times M_N = 1.55 \times 940 \text{ MeV} = 1455 \text{ MeV} , \tag{2.5}$$

assuming the same bag radius for both particles.

It was the merit of the MIT group to realize that Bogolioubov's bag model leads to violation of energy-momentum conservation at the bag surface, unless the internal pressure of the confined quarks is balanced at the boundary by some external pressure. The novel concept entering into the MIT-bag model [Ch74, Jo75, HK78] is that the true ground state of QCD, i.e. the QCD vacuum, is (partially?) destroyed inside the bag volume by the presence of quarks carrying colour. Whereas the true vacuum of QCD is a terribly complicated state only supporting excitations of colour singlet nature, the "vacuum" inside the bag allows for the presence of coloured particles and its physics can be treated by methods of perturbation theory. Often one speaks of the "perturbative vacuum". (For an extensive discussion of our present understanding of the true vacuum of QCD the reader is referred to the lectures of Shuryak [Sh84].)

For reasons of Lorentz invariance the exterior vacuum pressure (which really is a negative interior pressure) must be characterized by a scalar constant B, the bag constant, with dimensions of an energy density. The original fit of the MIT group to the spectrum of hadronic states [deG75] gave the value $B^{\frac{1}{4}} = 145$ MeV, but other values up to $B^{\frac{1}{4}} = 235$ MeV have been reported in the literature [Ha81]. The requirement of pressure balance at the surface leads to the quadratic boundary condition

$$-\tfrac{1}{2}n^{\mu}\partial_{\mu}\Big(\sum_i \bar{\psi}_i\psi_i\Big)\Big|_S = B\,, \tag{2.6}$$

where the sum runs over all quarks contained in the bag. One can show that for a spherical bag the condition (6) is equivalent to the requirement that the total energy contained in the bag volume be a minimum with respect to the bag radius R: $\partial M/\partial R=0$. Due to the scalar nature of the bag constant B, the difference between the vacua in the interior and the exterior of the bag contributes an energy that is proportional to the bag volume V:

$$M(R) = (\sum_i x_i)/R + (4\pi/3)BR^3 , \tag{2.7}$$

where $x_i=\omega_i R$. The original MIT value of B corresponds to a volume energy of 60 MeV/fm³. The equilibrium radius is found to be given by

$$R_0^4 = (\sum_i x_i)/(4\pi B) , \tag{2.8}$$

i.e. the radius of a hadronic bag is related to its mass through $M(R_0) = (16\pi/3)BR_0^3$. In short, M = 4BV, where V is the volume of the bag.

In order to obtain a quantitative description of the low energy part of the spectrum of hadronic states, in particular of the baryon octet and decuplet and of the meson nonets (with the notable exception of the pions) the bag model must be refined in several respects. The Casimir energy due to the zero-point fluctuations of gluon and quark fields inside the bag [Mi83] yields a contribution of the form $-Z_0/R$, as does the correction for spurious motion of the centre of mass [DJ80]. The original MIT fit used the values $Z_0=1.8$. The energy splitting between the baryon octet and decuplet, as well as between the nonets of pseudoscalar and vector mesons can be understood as being caused by the colour-magnetic interaction among the quarks. The interaction of magnetic type due to one-gluon exchange has the form

$$\Delta E_{mag} \propto \alpha_s/r \sum_{ij} \langle \lambda_i^a \lambda_j^a \sigma_i \sigma_j \rangle . \tag{2.9}$$

It is repulsive for parallel spins and attractive for antiparallel spin orientation. The original MIT fit [deG75] , see Fig. 4, was obtained with the coupling constant $\alpha_s=2.2$, but this is probably an exaggerated value forced by the neglect of other effects.

There is no obvious reason why a bag should not contain real gluons, although such states - called glueballs - have not been experimentally

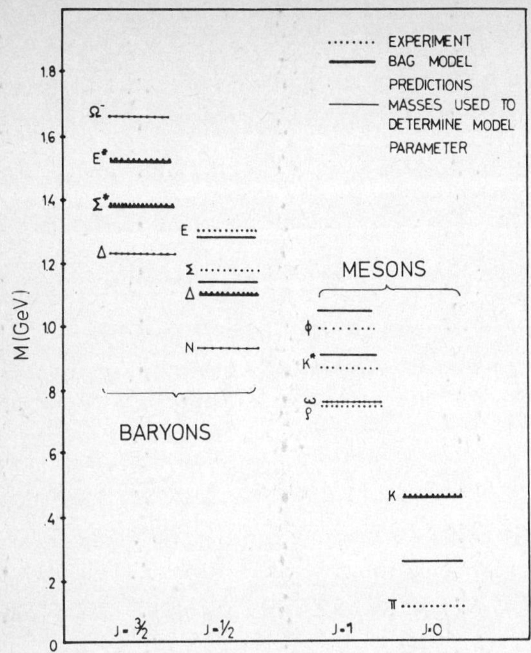

Fig.4: Bag model fit for the low-lying hadronic states [deG75]. The parameters are $b^{\frac{1}{4}}$ = 146 MeV, α_s = 2.2, Z_0 = 1.84, m_s = 279 MeV.

identified to date. The appropriate boundary conditions for the glue field are obtained from the requirement that the colour-electric field not be permitted to penetrate out of the bag into the "true vacuum". This is analogous to the boundary conditions in classical electrodynamis for a medium with dielectric constant $\varepsilon=0$ and magnetic permeability $\mu=\infty$:

$$\vec{n}\cdot\vec{E}^a = 0 , \quad \vec{n}\times\vec{H}^a = 0 , \quad \text{i.e.} \quad n_\mu F^a_{\mu\nu} = 0 . \tag{2.10}$$

By virtue of Gauss' theorem, an integration of the electric boundary condition over the full bag surface yields the result that the total colour charge contained in the bag volume must be zero, ensuring that all hadrons are colour singlets. For a spherical bag the stationary solutions of the free Yang-Mills equations (1,15) in their linearized form can be classified as transverse electric and transverse magnetic eigenmodes just as in electrodynamics. The boundary conditions for these modes are [Le79]:

$$\begin{aligned} &(\mathrm{TE}) \quad (d/dr)[r j_\ell(\omega r)]_{r=R} = 0 , \\ &(\mathrm{TM}) \quad j_\ell(\omega R) = 0 . \end{aligned} \tag{2.11}$$

Concerning hadronic phenomenology, the MIT-bag model has two important flaws. For one, the pion mass comes out too large in all versions of the model. The other objection is that the linear boundary condition (2) is not chirally invariant. In other words, the MIT-bag model grossly violates PCAC. There have been various attempts to cure these deficiencies by adding the pion field as an independent degree of freedom to the bag model [BR79, TMT80]. In these so-called "chiral bag" models the violation of quark axial vector current conservation at the bag boundary is partially balanced by the axial current from the pion cloud. In lowest order of the pion field the new boundary condition reads:

$$(\nabla^2 - m^2)\vec{\phi} = (i/2f)\,\bar{\psi}\gamma_5\vec{\tau}\psi\Delta_S\,, \tag{2.12}$$

where $\Delta_S = -n^\mu \partial_\mu \theta_V$ is the surface delta function and $f = 93$ MeV is the pion decay constant. The quarks thus act as a source of a pion cloud surrounding the bag.

There is the question of whether the pion should be allowed to penetrate into the interior of the bag where it would exist side by side with quarks and gluons. It seems that models which allow for such penetration, like the "cloudy bag" model [TMT80, TTM81] or the "Tel-Aviv-bag" model [KE83], are more successful in explaining the low energy properties of hadrons than those which do not, such as the "little brown bag" [BR79,BRV79]. For a review see [Th82].

The only relevance of this controversy for us is that we do not know whether pions can exist in the quark-gluon plasma or not. This is connected with the problem whether there are two separate phase transformations in QCD at high temperature, one leading to colour deconfinement, and the other leading to restoration of chiral symmetry, or whether in fact the two transitions coincide. In a chirally symmetric phase the pions, being the Goldstone bosons of chiral symmetry breaking, would not exist. As this issue has not been decided yet (see ch. 6 for some recent results), we shall forgo further discussion, because its effect on the equation of state of the quark-gluon plasma is not expected to be large (pions have three degrees of freedom, quarks and gluons together carry forty!).

Another deficiency of the MIT-bag model (and the standard chiral bag models for that matter) is that the surface is an idealized construction carrying no independent degrees of freedom. A possible cure fore this is provided by the so-called "soliton bag" models that derive more directly from Bogolioubov's original idea. In these models [Vi75,Ra76, HS76,FL77,FL78] the large mass of a quark outside the bag is generated by the coupling to a scalar field (in [RM76] the possibility of vector coupling was studied). Inside the bag the scalar field is dynamically suppressed by the presence of the quarks. In the case of Friedberg and Lee's model absolute confinement is ensured by the requirement that the chromo-dielectric constant of the vacuum vanishes when the scalar field is at full strength outside the bag volume. The soliton bag model has recently been very intensely studied by Goldflam and Wilets and collaborators [GW82]. Because it is much better suited for the description of the dynamical properties of the bag, such as vibrations of the bag surface, it may well be of use in further studies of the behaviour of giant quark-gluon plasma bags.

LECTURE 2

After this introduction we have a first look at the quark-gluon plasma, deriving its equation of state in the absence of interactions. We find that about half the energy resides in the quarks and half in the gluons, respectively. For the total energy density we obtain the convenient relation $\varepsilon = T^4$ (GeV/fm^3), when the temperature T is measured in units of 160 MeV. Utilizing the ideas of the bag model we derive a first estimate of the phase boundary between hadronic matter and the quark-gluon plasma in the T-μ-plane, where μ is the chemical potential of light quarks. We estimate the critical values $T_c \simeq 160$ MeV (for μ=0) and $\mu_c = 450$ MeV (for T=0). In view of the uncertainties the critical energy density is estimated to lie somewhere between $\frac{1}{2}$ and 2 GeV/fm^3.

We also discuss the prospect of reaching this value in a high-energy collision between nuclei, and mention several ideas how the formation of the quark-gluon plasma may occur in such a collision. We also investigate the rate of energy loss from the quark-gluon plasma by surface radiation of pions, where we find that it should leave sufficient time for the plasma to equilibrate and expand according to the laws of relativistic hydrodynamics.

We finally discuss the signals which have been proposed as experimental triggers for the quark-gluon plasma: photons and lepton pairs, strangeness, antinuclei and charge correlations. We compare the known particle yields of a proton-proton collision with that expected from the quark-gluon plasma.

3. EQUATION OF STATE

In order to investigate the properties of the quark-gluon plasma and how it may possibly be formed we have to derive its equation of state. This task is considerably simplified by the observation that the gluons and the light u and d quarks are essentially massless particles, at least on the scale of energies available in the hot plasma, i.e. 200 MeV. To obtain some first semi-quantitative insight we neglect all interactions among the constituents of the plasma (except for the assumption of thermalization which is a result of their presence). Let us first count the number of degrees of freedom associated with the constituents. Gluons carry colour and spin, and so do quarks which, in addition, come in two flavours u and d. We therefore find the following multiplicity:

$$\text{Gluons:} \quad N_g = 2(\text{spin}) \times 8(\text{colour}) = 16 ,$$

$$\text{Quarks:} \quad N_q = 2(\text{spin}) \times 3(\text{colour}) \times 2(\text{flavour}) = 12 \tag{3.1}$$

All we have to do now is to calculate the energy density residing in each degree of freedom. We begin with the gluons which, upon neglect of all interactions, form an ideal relativistic Bose gas of temperature $T = 1/\beta$:

$$\begin{aligned} \varepsilon_g &= \int(dp)\, p\, (e^{\beta p}-1)^{-1} = \\ &= 4\pi T^4/(2\pi)^3 \int_0^\infty x^3 dx (e^x-1)^{-1} = \pi^2 T^4/30. \end{aligned} \tag{3.2}$$

Here we have used the abbreviation (dp) for the density of states $d^3p/(2\pi)^3$, and we have rescaled the momentum $x=\beta p$.

For the quarks and antiquarks we have to introduce a chemical potential μ, because there will be in general a surplus of quarks over antiquarks in the quark-gluon plasma. It may be that $\mu=0$ in special situations, but if the quark-gluon plasma is produced in nuclear collisions, a net baryon number excess must be expected. At zero temperature, the meaning of μ is the energy required to add another quark to the plasma. In other words, μ is the Fermi energy of the quarks. Since no antiquarks are present at T=0, the energy necessary to add an antiquark is zero. This does not imply $\mu=0$, however, because the additional antiquark may annihilate with one of the quarks from the top of the Fermi sea and release the energy μ. The chemical potential of the antiquarks must therefore be chosen as $-\mu$.

It turns out that the energy density residing in the quarks alone, or the antiquarks alone, cannot be calculated analytically in the general case $\mu, T\neq 0$. Miraculously, however, the sum of both yields a simple analytical formula as we shall see now [Ch78,EGR80]. The energy density per degree of freedom carried by quarks is

$$\varepsilon_q = \int (dp)\, p\, [e^{\beta(p-\mu)}+1]^{-1} = T^4/2\pi^2 \int_{-\beta\mu}^{\infty} dx\, (x+\beta\mu)^3\, (e^x+1)^{-1} . \tag{3.3}$$

where we have scaled and shifted the momentum variable according to $x = \beta(p-\mu)$. The only change occurring in the same expression for antiquarks is the replacement $\mu\to(-\mu)$, which is compensated by the variable change $x = \beta(p+\mu)$:

$$\varepsilon_{\bar{q}} = \int (dp)\, p\, [e^{\beta(p+\mu)}+1]^{-1} = T^4/2\pi^2 \int_{\beta\mu}^{\infty} dx\, (x-\beta\mu)^3\, (e^x+1)^{-1} . \tag{3.4}$$

Splitting the last integral in the way $\int_{\beta\mu}^{\infty} = \int_0^{\infty} - \int_0^{\beta\mu}$, and making the substitution $x\to(-x)$ in the second integral, the latter becomes

$$\int_{-\beta\mu}^{0} dx\, (x+\beta\mu)^3\, [1-(e^x+1)^{-1}] . \tag{3.5}$$

Combining the expression for the quark energy density, the first part of the integral in eq.(4) and eq.(5) we obtain:

$$\varepsilon_q + \varepsilon_{\bar{q}} = T^4/2\pi^2 \{{}_0\int^\infty dx[(x+\beta\mu)^3+(x-\beta\mu)^3](e^x+1)^{-1} + {}_{-\beta\mu}\int^0 dx(x+\beta\mu)^3\}$$
$$= 7\pi^2T^4/120 + \mu^2T^2/4 + \mu^4/8\pi^2 . \tag{3.6}$$

This is a very simple expression, indeed, which relies on the assumption of vanishing rest mass of the quarks.

In order to get a feeling for the range of energy densities involved in quark-gluon plasma physics, let as first consider the case of baryon number-symmetric quark-gluon matter, i.e. the case $\mu=0$. Multiplying the expressions (2) and (6) by the number of respective degrees of freedom given in eq.(1) we find:

$$\varepsilon = 16\varepsilon_g + 12(\varepsilon_q+\varepsilon_{\bar{q}}) = 37\pi^2T^4/30 = (T/160\text{MeV})^4 \text{ GeV/fm}^3 . \tag{3.7}$$

This number must be compared with the density in nuclear matter $\varepsilon_{nuc} \simeq 125$ MeV/fm^3 and the energy density in the interior of a nucleon which is four times the MIT-bag constant: $\varepsilon_N \simeq 4B \simeq 300\text{-}500$ MeV/fm^3. We conclude that the energy density in the hot quark-gluon plasma must exceed the energy density inside an individual hadron by at least a factor two.

It is interesting to note from eq.(7) that at $\mu=0$ quarks and gluons contribute about the same amount to the energy density of the quark-gluon plasma. This is not trivial, for had we taken the group SU(2) as the colour group, where there are two quark colours and three gluon colours, the contribution from gluons would have been only half of that of the quarks. If we include the effect of the chemical potential of the quarks, e.g. by considering quark matter with μ approximately equal to one-third of the nucleon mass, i.e. $\mu \simeq 300$ MeV, the quark contribution to the total energy density at T = 160 MeV triples, yielding $\varepsilon \simeq 2$ GeV/fm^3.

In order to determine the value of the quark chemical potential μ in a given physical situation one has to know its relation to the baryonic density n_b. n_b is one third of the difference between the density of quarks and the density of antiquarks. In analogy to eqs.(3.4) we have per degree of freedom:

$$n_q = \int(dp)\ [e^{\beta(p-\mu)}+1]^{-1}$$

$$n_{\bar{q}} = \int(dp)\ [e^{\beta(p+\mu)}+1]^{-1} \quad . \tag{3.8}$$

Utilizing the same tricks as employed in the evaluation of the energy density, we obtain

$$\begin{aligned} n_q - n_{\bar{q}} &= T^3/2\pi^2\ \{{}_0\int^\infty dx[(x+\beta\mu)^2-(x-\beta\mu)^2](e^x+1)^{-1} + {}_{-\beta\mu}\int^0 dx(x+\beta\mu)^2\} = \\ &= \mu T^2/6 + \mu^3/6\pi^2 \quad . \end{aligned} \tag{3.9}$$

Multiplying by the number of degrees of freedom, i.e. by 12, and dividing by 3 we find

$$n_b = 4(n_q - n_{\bar{q}}) = 2\mu T^2/3 + 2\mu^3/3\pi^2 \quad . \tag{3.10}$$

At zero temperature this is the well-known expression for the degenerate Fermi gas, but for high temperature the first term causes the chemical potential to drop like T^{-2} if the baryon density is kept constant. Incidentally, we find the relation

$$n_b = 4/3\ \partial\varepsilon/\partial\mu \ , \tag{3.11}$$

which has a fundamental thermodynamic origin and remains valid if interactions among the particles are taken into account (see ch. 8).

The pressure P and the entropy density s of the quark-gluon plasma may be calculated in a similar way, with the result

$$P = 1/3\ \varepsilon \quad , \qquad s = 1/3\ \partial\varepsilon/\partial T \quad . \tag{3.12}$$

These results which are generally true for systems composed of massless particles, also remain valid when interactions are included. How this is done will be discussed in detail in ch. 8. The result is a modification of the factors in front of the individual terms in eqs. (6) and (9). In lowest order of the QCD coupling constant α_s the expression is given in eq. (5) of ch. 8.

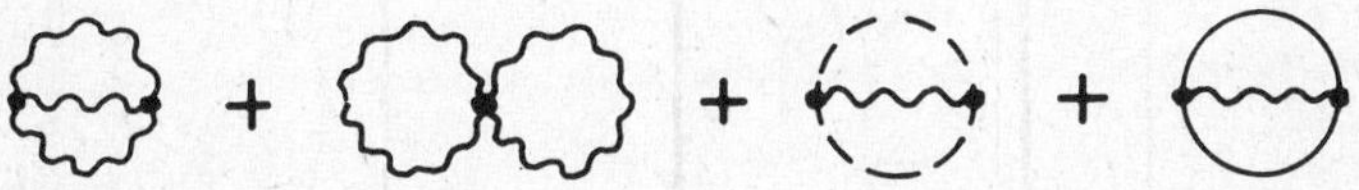

Fig.5: Feynman diagrams contributing to the equation of state of the quark-gluon plasma in order α_s. Wavy lines represent gluons, solid lines quarks, and dashed lines denote the ghost subtractions of unphysical degrees of freedom.

We can get an idea of the range of stability of the quark-gluon plasma phase if we inquire under which conditions the internal pressure from the coloured quarks and gluons can balance the external vacuum pressure B as introduced in the MIT bag model. We would say that the plasma phase is expected to be stable if $P = \varepsilon/3 \geq B$, equality giving the boundary of the stability region. Calling the values of μ and T on the boundary *critical* values, μ_c and T_c, we obtain a relation between them:

$$B = \pi^2 T_c^4 [(37/90 - 11\alpha_s/9\pi) + (1 - 2\alpha_s/\pi)(x_c^2 + \tfrac{1}{2}x_c^4)] \quad , \tag{3.13}$$

where we put $\mu_c = x_c \pi T_c$.

The phase boundary is shown in Fig.6 for the two values $\alpha_s = 0, \frac{1}{2}$ in units of $B^{\frac{1}{4}}$. In terms of the variables μ and T the uncertainty expressed by those

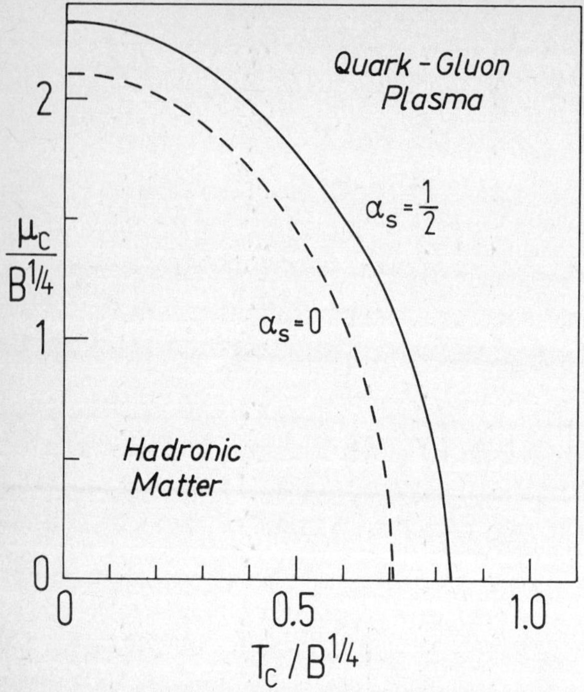

Fig.6: Phase diagram for hadronic matter and quark-gluon plasma in the μ-T plane. Solid line: with interaction, dashed line: without interaction.

two choices makes only a 10-15 percent difference. The relevant quantity for experimental considerations, however, is the energy density which goes like the fourth power of μ and/or T. The small uncertainty in the values of μ_c and T_c at the boundary of the stability region means that the critical energy density is uncertain by a factor of two, easily. In view of eq. (7) we conclude that our best estimate based on these considerations is

$$\varepsilon_c = \tfrac{1}{2} - 2 \text{ GeV/fm}^3 \quad . \tag{3.14}$$

A more reliable number can only come from full-scale numerical evaluation of the equation of state of QCD based on Monte-Carlo calculation on the lattice (see ch. 6). Not before these include quarks and virtual quark-antiquark excitations will it be possible to quote a better value for ε_c and to say how well-defined the change of phase from hadronic matter to the quark-gluon plasma really is.

But let us continue and ask how we can imagine the quark-gluon plasma to be produced. To be sure, it must have existed during the very early phase of the expansion of the universe. The average temperature of the universe today is 2.7 K. With the help of Einstein's equations of general relativity it is possible to extrapolate backward in time to the instant when the temperature reached 200 MeV. Since the expansion law in the standard cosmological model (see e.g. [We72]) is of the form $R(t), T(t) \propto t^{\frac{1}{2}}$, the transition from quark-gluon plasma to hadronic matter must have occurred about 20 μsec after the big bang, when the curvature radius of the universe was approximately 1/50 light year.

During the phase transition there was a short period of exponential growth of the radius of the universe due to the change between two different vacuum states ("inflationary universe" [Gu81]). Under the most likely and reasonable assumptions, the time-scale for the expansion of the universe (10^{-5}s) was very much longer than the time required for the phase transition to occur locally (10^{-23}s). Hence, the phase change probably proceeded adiabatically, leaving no presently noticeable imprint on the evolution of the universe (see e.g. [vH84]).

However, if the transition between the quark-gluon plasma phase and the hadronic phase was sufficiently impeded by some mechanism, leading to extended supercooling of the deconfined phase, density fluctuations of considerable size may have been created. The phase transition may then be of importance for our understanding of inhomogeneities in our universe, such as black holes, galaxies, etc. [O181,CS82,KT82,Su82,LD83,Ho83, DK84,S084]. An even more spectacular scenario is that of Witten who speculates that the 'invisible' mass in the universe may be composed of left-over strange quark matter [Wi84].

We cannot repeat the creation of the universe in order to look for details of the phase transition. The only promising way to investigate the properties of the quark-gluon plasma by means of laboratory experiments is the study of central collisions between heavy nuclei at very high energy [AKM80]. Our expectations as to the nature of such events are derived from studies of nuclear collisions up to 2.1 GeV per nucleon (at the Bevatron in Berkeley) for medium heavy nuclei and up to 4 GeV per nucleon (at Dubna) for light nuclei, as well as from high-energy hadron-nucleus interactions [BG84,CK84,Hw84,Wo84]. Some relevant data are also available from cosmic ray experiments [Yo82,Bu83,Bu84].

The conclusions drawn from these studies are the following. At energies below about 6 GeV per nucleon a heavy target nucleus, like uranium, is able to stop the projectile [DGS85]. In the course of this process most of the kinetic energy carried by the incident nucleus is thermalized, and some energy is converted into compression energy. Calculations in the framework of the hydrodynamical model [SGB81] and the intranuclear cascade model indicate that nuclear matter may be compressed up to ten times its normal density.

In order to obtain a very crude estimate of the energy density that can be attained in such a collision, we follow an argument of Goldhaber and Gyulassy [Go78,Gy83b]. Let us consider a collision of two nuclei with kinetic energy per nucleon E/A in the lab system. In the center-of-mass system both nuclei are lorentz-contracted by a factor γ with

$$\gamma^2 \cong 1 + \tfrac{1}{2}E/(AM_N)\ , \quad \gamma^2 \approx 1 + \frac{1}{2}\frac{K.E_{lab}}{M}\left(1+\frac{K.E_{lab}}{2M}\right) \tag{4.1}$$

where M_N is the nucleon mass. The total energy per nucleon in the c.m. system is γM_N. If both nuclei collide with each other and come to a stop when they overlap completely, the energy density, before expansion starts, is (see Fig. 7):

I get O $\xrightarrow{p_{lab}}$ O Transform To CM O $\xrightarrow{p_{cm}}$ $\xleftarrow{p_{cm}}$ O $p_{cm} = \frac{1}{2}p_{lab}$

(E_{lab}, p_{lab}) $(M, 0)$ (E'_{cm}, p_{cm}) $(E'_{cm}, -p_{cm})$

$p_{lab} = \sqrt{E_{lab}^2 - M^2}$ $E_{lab} = K.E. + M^2$ $E'_{cm} = \gamma M = \sqrt{M^2 + p_{cm}^2}$

$$\gamma^2 M^2 = M^2 + \frac{1}{4}(E_{lab}^2 - M^2) = M^2 + \frac{1}{4}(K.E._{lab})\frac{(K.E. + 2M)}{M^2}$$

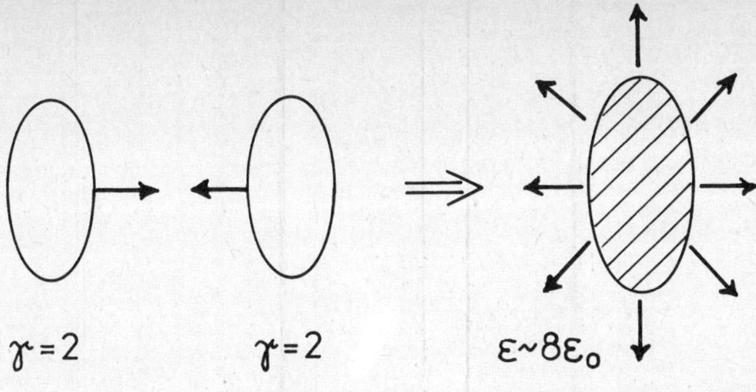

Fig.7: Collision between two nuclei below 10 GeV/nucleon as seen from the c.m. system. The Lorentz contracted nuclei collide and come to a stop.

$$\varepsilon/\varepsilon_0 \;=\; 2\gamma^2 \;=\; 2 + E/(aM_N) \;. \tag{4.2}$$

Here $\varepsilon_0 \simeq 125$ MeV/fm^3 is the normal energy density of ground state nuclear matter. Since this scenario is a reasonable expectation up to energies of 5-10 MeV per nucleon, energy densities up to 1-1.5 GeV/fm^3 may be reached. According to our estimates in the previous chapter the energy density of quark matter is of the order of 1-2 GeV/fm^3, the critical density may just be reached in such collisions. At the same time the baryonic density would be $n \simeq 2\gamma n_0$, reaching up to 0.6/fm^3. According to eq. (3.10) this corresponds to a chemical potential of 200-250 MeV for the quarks at a temperature of 160 MeV. Matter under these conditions would contain many quarks and gluons but few antiquarks.

At energies higher than about 10 GeV per nucleon the nuclei will no longer be able to stop each other's motion, and at much higher energy (30-100 GeV/nucleon or more) they are expected to be essentially transparent. At first sight this may be surprising, because nucleon-nucleon cross-sections saturate at a value of about 40 mb, implying a mean free path of only 1.5-2 fm for a nucleon in nuclear matter even at very high incident energy. However, when a nucleon has scattered it takes a time of about $\tau_0 \simeq 1$ fm/c in its own rest frame until the fragments have lost their correlations and are able to scatter again as independent particles. Due to time dilatation ("longitudinal growth" [Ni81]) the nucleon and its fragments have then travelled a distance

$$R = \tau_0 cE/(AM_N) \tag{4.3}$$

in the laboratory system. At sufficiently high energy, this point of materialization will be far outside the target nucleus, therefore one speaks of the "inside-outside" cascade (see Fig. 8).

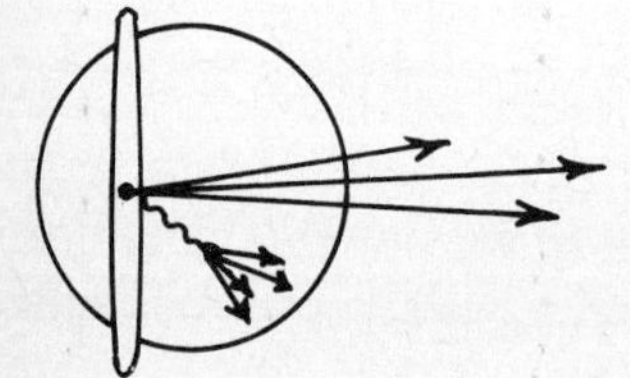

Fig.8: At very high energy time dilatation causes the projectile fragments to fly a long way before they materialize on the mass shell. An inside-outside cascade develops.

After the passage of the highly Lorentz contracted projectile the target nucleus remains in a violently excited state. The target-like nucleon fragments have time to rescatter several times off one another and assume a state close to thermal equilibrium. Due to the many mesons created by collisions between target and projectile nucleons (pions up to about 2 GeV/c momentum become trapped in the target nucleus) the energy density of the nuclear matter in the target fragmentation region is expected to be very high, about 1 GeV/fm^3 [KRR83,St84]. The same description applies to the matter in the projectile nucleus, if looked at from a comoving reference system.

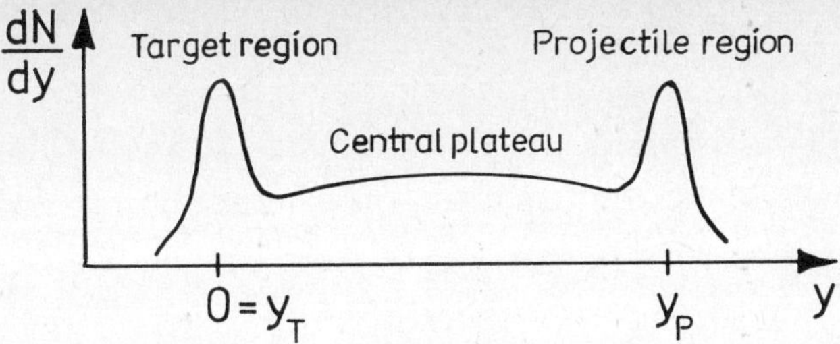

Fig.9: At very high energy a central plateau region opens between the target and projectile fragmentation regions. Within the central plateau the average multiplicity of particles per interval of rapidity is approximately constant.

Particle collisions at high energy are conveniently represented in the rapidity variable $y = \text{artanh}(v_{\|}/c)$, because a Lorentz boost in the direction of the beam corresponds to a constant shift in rapidity. The distribution of particles after the collision is sketched in Fig. 9. There are the highly excited baryon-rich fragmentation regions in the vicinity of the initial rapidity of the target, y_T, and of the projectile, y_P. At sufficiently high collision energy these two regions are expected to be well separated with a central region extending between them, where the average multiplicity per rapidity interval, dN/dy, is approximately flat. As the overwhelming majority of fragments produced in the collision are mesons, the central region will contain few baryons, i.e. the chemical potential is $\mu \simeq 0$. If a quark-gluon plasma is produced in this region [Ka82,McL82] it will contain as many quarks as antiquarks (Fig. 10).

The energy density in the central region is also expected to be quite high (about 2 GeV/fm^3 [KRR83]) with a considerable amount of fluctuations [ELM83]. The central region as well as the fragmentation regions are promising candidates for an experimental search for the quark-gluon plasma. At ultra-high, cosmic ray energies the central region is probably

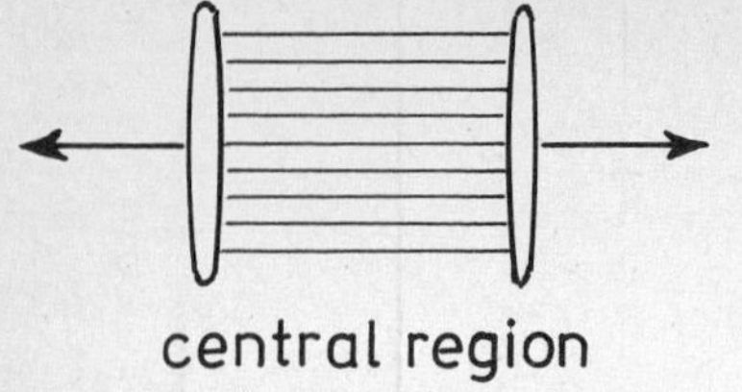

Fig.10: The central region opening up after the collision between target and projectile contains mostly mesonic matter, with an equal amount of quarks and antiquarks.

preferred, but at (laboratory) energies up to 100 GeV per nucleon this is not necessarily so. The space-time evolution of the quark-gluon plasma in the framework of the inside-outside cascade and the hydrodynamical model will be discussed in more detail in ch.11.

Now that we have discussed the possible formation of the quark-gluon plasma in nuclear collisions at high energy, let us speculate how it may decay. The two mechanisms that immediately come to mind are: expansion until the density drops below the threhold of stability for quark matter, or cooling by emission of particles (mainly pions) from the surface of the fireball (see Fig. 11). While mostly (isentropic) expansion has been considered primarily responsible for cooling [Ka82,Ka82a,McL82, Bj83], Danos and Rafelski [DR83,RD83] claimed that surface radiation of pions is the dominant process. On the other hand, Banerjee, Glendenning and Matsui [BGM83] found that meson evaporation is strongly suppressed in the framework of the flux-tube model [CNN79,GM83,Va83].

Let us try to estimate the energy loss from a region of quark-gluon plasma in empty space due to pion radiation. Normally the plasma will be surrounded by other hot nuclear matter which will cause some reheating. In neglecting this backflow of energy we will get an upper estimate for the energy loss. The simplest model is obtained by calculating the thermal flow of pions from the surface of the quark-gluon plasma. According to Planck's law the radiated energy per unit time and unit surface area is

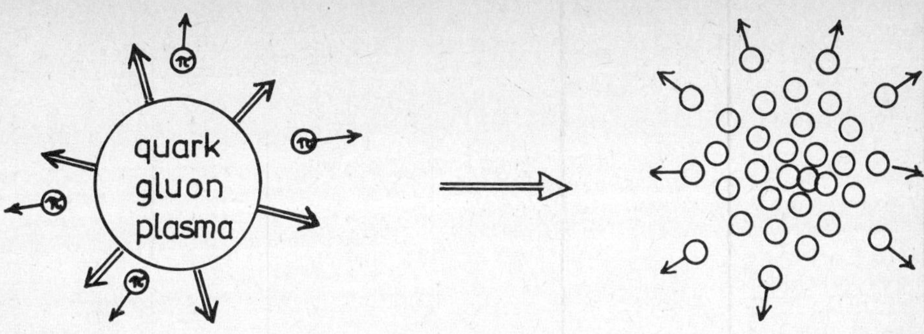

Fig.11: Expansion of the quark-gluon plasma fireball. Pion evaporation competes with (isentropic) hydrodynamic expansion.

$$dE/(dtd^2x) = 3 \int d^3k/(2\pi)^3 \; k_0(e^{\beta k_0}-1)^{-1} \; (\vec{n}\cdot\vec{k}/k_0)\theta(\vec{n}\cdot\vec{k}) \tag{4.4}$$

where n is the surface normal vector and the factor 3 counts the isospin degrees of freedom. Carrying out the angular integrations we find

$$\begin{aligned} dE/(dtd^2x) &= (3/8\pi^2) \int_m^\infty k^3 dk \sum_n \exp[-n\beta(k^2+m_\pi{}^2)^{\frac{1}{2}}] = \\ &= 3T^4/4\pi^2 \sum_n n^{-4} \exp(-n\beta m_\pi)[(n\beta m_\pi)^2+3n\beta m_\pi+3] \; . \end{aligned} \tag{4.5}$$

For $T \geq m_\pi$ the coefficient of T^4 turns out to be a slowly varying function of the parameter βm_π that is well represented by

$$dE/(dtd^2x) \simeq 0.2\; T^4 \; . \tag{4.6}$$

In order to obtain a characteristic cooling time we integrate (6) over the surface, assuming a spherical volume of radius R, and divide by the total energy contained in the quark-gluon plasma. With the equation of state (3.7) we find

$$\tau = E(dE/dt)^{-1} = (37\pi^2T^4/30 \times 4\pi R^3/3)\,(T^4/5 \times 4\pi R^2)^{-1} \simeq 20\ R\ . \tag{4.7}$$

A plasma droplet of 2 fm radius would lose half its energy within a period of 20 fm/c or 6×10^{-23}s. This is a very long time by strong interaction standards, and we would expect that expansion of the plasma fireball under its own internal pressure is a much more efficient cooling mechanism.

But let us try to understand the origin of the large numerical factor 20 in eq.(7). It can be related to the difference of the number of degrees of freedom in the plasma (40, counting quarks and gluons) and those available for radiation (three). For every ten coloured particles hitting the surface from inside only one pion is emitted, on the average. Is this number, which arises from plain statistical arguments, reasonable in the context of a dynamical radiation model?

There is only one way to answer this question: by computing the rate of pion emission from particles hitting the plasma bag boundary. To do so we may use one of the chiral bag models which contain a surface coupling between quarks and the pion field, as discussed in ch.2. We first note that gluons do not create pions in the framework of such a model, because their axial current vanishes. This argument eliminates half of the internal degrees of freedom from contributing to pion production. There are three different processes: (a) a quark can bounce off the surface losing some energy which is carried away by the pion, (b) an antiquark can do the same, and (c) a quark and an antiquark can meet at the surface and annihilate into a pion. In the first two cases the reflected (anti-) quark has to find an unoccupied final state because of the Pauli principle (see Fig. 12).

The number of pions radiated from the surface is then

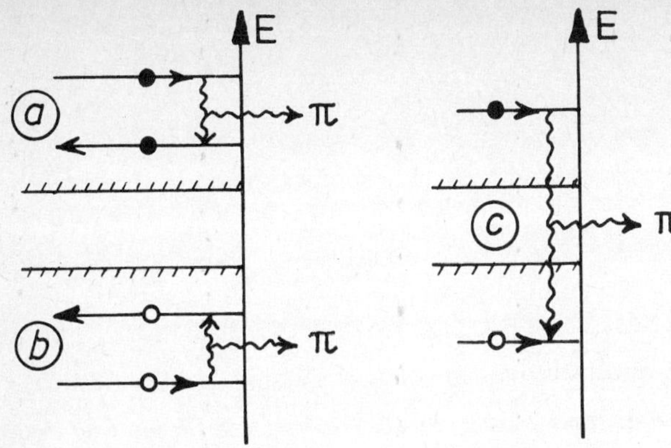

Fig.12: Three processes contribute to the creation of pions on the surface of a quark-gluon plasma bag: (a,b) a quark or antiquark loses kinetic energy while bouncing off the surface, (c) a quark and an antiquark annihilate at the surface.

$$N_{\pi} = \int(dpdp'dk)\ \rho(p)[1-\rho(p')]\ (\Sigma\ |S_{fi}(p,p',k)|^2)\ , \tag{4.8}$$

where p, p' and k denote the momenta of the impinging, the reflected quark and the pion, respectively (see Fig. 13). $\rho(p)$ and $\rho(p')$ describe the occupation probabilities of the initial and final states as given by eqs.(3.8). The sum runs over the discrete quantum numbers of the initial and final states including the pion, i.e. spin, colour and isospin.

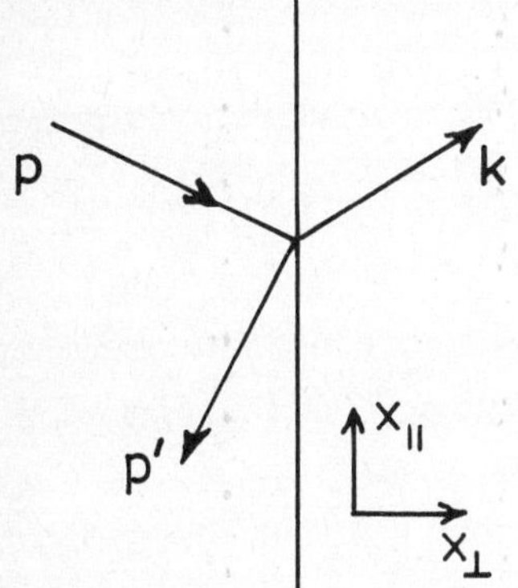

Fig.13: Notations for the momentum variables used in evaluation of the pion creation amplitude $S_{fi}(p,p',k)$.

The S-matrix element is obtained by evaluating the quark-pion interaction Lagrangian (2.12)

$$L_{q\pi} = (i/2f)\, \bar{\psi}\gamma_5 \vec{\tau}\cdot\vec{\phi}\psi\, \Delta_S \tag{4.9}$$

with plane waves for the ingoing and outgoing particles. There is one small technical problem with this. At the bag surface the quark wavefunctions must satisfy the linear boundary condition (2.2), i.e. $i(\gamma n)\psi=\psi$. Since the quark-pion interaction was derived from the violation of PCAC due to this boundary condition and involves a surface delta function, the wavefunctions should be properly constructed as standing waves that obey the boundary condition. As we need the quark wavefunctions only at the surface, there is an elegant way out. One can show that it is sufficient to integrate over all plane waves if one projects out the part satisfying the wrong boundary condition, i.e. $i(\gamma n)\psi=-\psi$. The required projection operator is $P_n = \frac{1}{2}(1+i\gamma^\mu n_\mu)$,since one easily verifies that

$$i\gamma^\mu n_\mu (P_n\psi) = P_n\psi \tag{4.10}$$

by remembering that $(\gamma n)^2=-1$. Note that this trick works only at the surface but not inside the bag.

The summed, squared S-matrix element may be evaluated by standard trace techniques for Feynman diagrams with the result

$$\begin{aligned}\Sigma\, |S_{fi}|^2 = 9\pi^3/(8f^2 p_0 p_0' k_0)\, (p_0 p_0' - p_\| p_\|') \times \\ \times\, \delta(p_0 - p_0' - k_0)\,\delta^2(p_\| - p_\|' - k_\|) \int dt d^2x \;.\end{aligned} \tag{4.11}$$

Here the index "$\|$" denotes the components of p or x parallel to the surface, which is idealized as an infinite plane. The delta function implying conservation of the normal momentum component is missing as a result of the surface delta function in the interaction Lagrangian. This is certainly an oversimplification which is incompatible with the existence of a pion formfactor and should be relaxed in future calculations. For the annihilation process (c) the sign of p' in the arguments of the delta functions in eq.(11) must be inverted.

The energy radiated per unit time and unit surface area is obtained by dividing eq.(8) by $dtd^2x_\parallel$ and adding a factor $k_\perp\theta(k_\perp)$ under the integrals. Because of our projection trick there is an additional factor 2 in each of the integrations over p and p'. Of the nine integrations in eq.(8) four may be carried out analytically. The remaining five-dimensional integral can be evaluated by Monte-Carlo methods after a factor T^6 has been extracted. We obtain the final expression

$$dE/(dtd^2x) = 9T^6/(8\pi^4 f^2)\ [I_{qq} + I_{\bar{q}\bar{q}} + I_{q\bar{q}}]\ , \tag{4.12}$$

where the dimensionless integrals I_ν depend on the parameters $\beta\mu$ and βm_π. The numerical evaluation of the integrals has been performed by Schnabel and Rafelski [SR84] and by Müller and Eisenberg [ME85] using Monte-Carlo methods. For temperatures above T = 150 MeV and $\mu = 0$ it was found that the finite rest mass of the pion has only a very small influence. The result is well represented by the formula

$$dE/(dtd^2x) \simeq 27T^6/(16\pi^4 f^2) = 0.02(T^6/f^2)\ . \tag{4.13}$$

The expression (13) grows more rapidly with rising temperature than eq.(6), resulting in faster cooling at higher temperatures. Since the T^6-dependence has its origin in the surface delta interaction between pions and quarks, which is an idealization, we must expect that eq.(13) ceases to be valid at very high temperatures.

The radiated power calculated from eq.(13) is compared with thermal pion emission in Fig. 14. As can be seen, the pion radiation rate predicted by the chiral bag model falls below the thermal emission rate up to temperatures of 300 MeV. With increasing quark chemical potential μ the pion

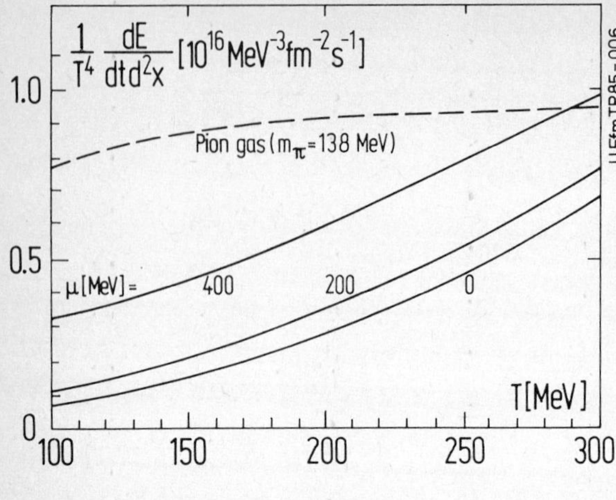

Fig.14: Power (divided by T^4) radiated by pions per unit surface area, for different values of the quark chemical potential (from [SR84]).

radiation grows more intense, but remains smaller than the thermal yield for values of T and μ that can be reasonably expected in a nuclear collision. The spectral distribution of pions from the three different processes illustrated in Fig.12 is shown in Fig.15.

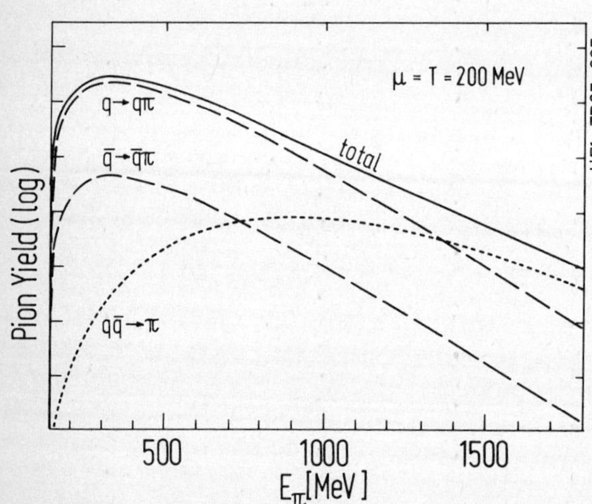

Fig.15: Energy spectrum of emitted pions in the chiral bag model. The contributions from the three processes of Fig.12 are shown separately.

The influence of the finite size of the quark-gluon plasma fireball (discreteness of quark energy levels) on the pion emission was investigated in

[ME85]. It turns out that the relative contributions from the three production processes shown in Fig. 12 are somewhat different for a bag of finite size, but the total pion yield is not much changed. However, the spectral distribution of the emitted pions depends very sensitively on the bag size, throwing some doubt on the validity of the surface coupling (9) between quarks and gluons. Still, our considerations lead us to the conclusion that cooling of the quark-gluon plasma by pion radiation is probably not a major channel of energy loss.

We finally mention that the chiral bag model coupling (9) between quarks and pions can be extended to include other pseudoscalar mesons (kaons, ϕ-mesons) in the framework of the chiral flavour SU(3)×SU(3) model of Gell-Mann and Lévy [GL60,Le67]. The chiral SU(3)×SU(3) group is still a fairly good symmetry, as the experimental values of the kaon and pion decay constants are similar: $f_K/f_\pi \simeq 1.25$. By considering also strange quarks and all processes leading to kaon formation, e.g. $s \to Kq$, $q\bar{s} \to K$, etc., one finds that the ratio between kaon and pion radiation at $T = 160$ MeV and $\mu = 0$ is quite large:

$$N_K/N_\pi \simeq f_\pi^2/f_K^2 \simeq 0.6 . \tag{4.14}$$

Although this result should be taken with appropriate caution, it indicates that the formation of a quark-gluon plasma is expected to be accompanied by an anomalously high K/π ratio. A similar prediction has been made on the basis of entropy arguments [GR84]. The importance of an enhanced strangeness abundance as a characteristic signal for the quark-gluon plasma is further discussed in the next chapter.

Now that we have seen that surface evaporation is not the dominant mechanism by which the quark-gluon plasma formed in nuclear collisions breaks up into hadrons, we have to ask: how does the quark-gluon plasma hadronize when it has cooled below the critical temperature T_c? One possibility is that the rearrangement from free quarks to hadrons proceeds gradually, with appropriate clusters of quarks already forming above T_c in the plasma phase [CCR84]. On the other hand, the rapid liberation of new degrees of freedom observed in Monte-Carlo simulations of lattice QCD (see ch.6) points in a different direction. Most likely, the structure of strongly interacting matter is very much different on both sides of the phase transition.

The phase change back to hadronic matter then implies a massive rearrangement of the internal structure of the quark-gluon plasma. A model for this hadronization process has recently been proposed by van Hove [vH85]. He assumes that a network of colour strings forms in the plasma when the temperature falls below T_c. The tension of these strings may bring the expansion of the quark-gluon plasma close to a stop. Occasionally one of the strings breaks by quark pair production leading to subdivision of the fireball into smaller droplets as shown in Fig.16. This sequential fissioning process [Va83] may be responsible for the large fluctuations seen in the multiplicity distributions in rapidity space, dN/dy, of fragments in cosmic ray events [Gy83b].

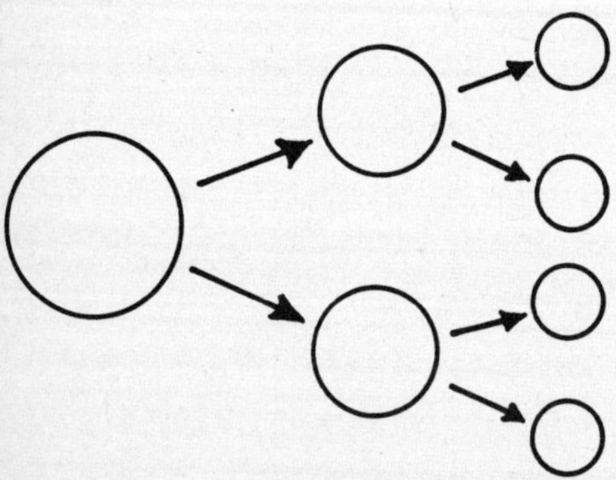

Fig.16: Sequential fission of the fireball is a possible mechanism for the break-up of the quark-gluon plasma.

The hot fireball will remain in the quark-gluon plasma phase only for a very short time (typically $\tau_p \simeq 5\text{-}10$ fm/c = $1.5\text{-}3\times10^{-23}$ s are expected). After this time its temperature has dropped below the critical temperature T_c due to expansion and radiation of particles from the surface. Below T_c the fireball matter will return to the hadronic phase, with a possible intermediate period of supercooling. Initially, the density of the hadronic phase will still be very high and only drop during further expansion. Until the density has become so low that the average inter-particle spacing is larger than the range of strong interactions, the hadrons in the fireball still interact violently. If the fireball would exhibit any characteristic *kinematical* signature during the time it spends in the quark-gluon plasma phase (as has been suggested by Shuryak and Zhirov [SZ80] and by van Hove [vH82]), this signature would run a good chance of being destroyed by the hadronic final state interactions. If one were to rely on an analysis of the momentum space distribution of final hadrons to prove the temporary existence of the quark-gluon plasma, the experiments might be quite inconclusive.

An exceptional case may be the scenario discussed recently by Gyulassy et al. [Gy84] and by van Hove [vH83]. They assume massive supercooling of the plasma phase during the expansion. The subsequent transition to the hadronic phase proceeds like an explosion producing shock waves that should be clearly visible in the transverse momentum distribution of the emitted hadrons (an extended discussion of the possible signatures associated with kinematical variables is found in [Gy84a]).

Any analysis promising success must be based on the observation of properties that are not affected by the final state interactions. Two such observables have been discussed in some detail: (a) particles that do not interact strongly, and (b) quantum numbers that remain unchanged by strong interactions.

Before we discuss the various proposed signatures in detail, it is worthwhile to draw attention to a characteristic property of the quark-gluon plasma. The number of degrees of freedom accessible to the coloured particles in the quark-gluon plasma is much higher than that available to the particles which constitute the hadronic phase. This is a direct consequence of the local deconfinement of colour in the quark-gluon plasma. As a result, the energy per particle is much lower than expected in a hadronic gas, because the total available energy is distributed over more degrees of freedom.

Almost all suggested signatures of the quark-gluon plasma are, therefore, based on the much *enhanced abundance* of certain particles at relatively *low energies*, typically in the range $T < E < 2T$ ($\simeq$ 150 MeV - 400 MeV). However, this is also the source of a major difficulty: since low energy phenomena in a violent collision are usually very involved and hard to calculate, a quite substantial, 'model-independent' enhancement due to the quark-gluon plasma is required lest a signal is swamped by fluctuations and not well understood background processes.

(a) Photons and lepton pairs:

Particles that interact only electromagnetically are produced by plasma processes with sufficient abundance to be experimentally detectable. Photons, in principle, are very nice [Ki80,CDH83], but in this case one has to deal with formidable background problems because of hadronic decays into photons, most notably of the π^0 (into 2γ) and the η (into 2γ or $3\pi^0$). In principle, photons from these decays can be distinguished by invariant mass analysis, but this seems very difficult if 100 neutral pions or more are produced in the nuclear reaction. On the other hand, photon production occurs throughout the fireball volume, whereas pions are predominantly radiated off the surface. For collisions of very heavy nuclei (e.g. uranium) this may result in a fairly large ratio of $n_\gamma/n_\pi \simeq 0.2$ [CDH83].

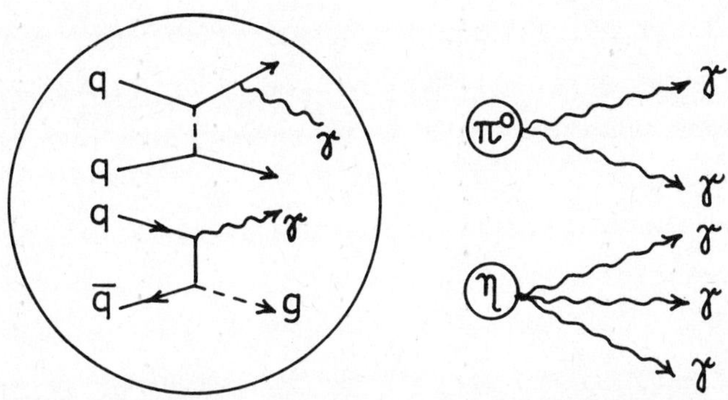

Fig.17: High-energy photons are produced in quark-quark and quark-antiquark collisions. However, electromagnetic decay of neutral pions and etas produces a large background of radiation.

The observation of high-energy lepton pairs, in particular e^+e^- but also $\mu^+\mu^-$, seems to be more promising [Fe76,DG81,KM81,Ka82]. The dominant background process in this case is $\pi^+\pi^-$- annihilation with subsequent lepton pair production (see Fig.18). Fortunately, this proceeds mainly through the ρ-meson channel and the invariant mass of the lepton pair is therefore concentrated in the region around 770±100 MeV. Calculations by Domokos and Goldman [DG81,Do83] and by Chin [Ch82] indicate that lepton pairs from the quark-gluon plasma should be quite dominant in the lower mass region between 300 and 500 MeV (see Fig. 19). The size of their contribution could be a very sensitive measure of the temperature initially reached in the plasma phase, and also of the effective quark mass [Ka84] inside the quark-gluon plasma (but see [Gy84a] for critical comments).

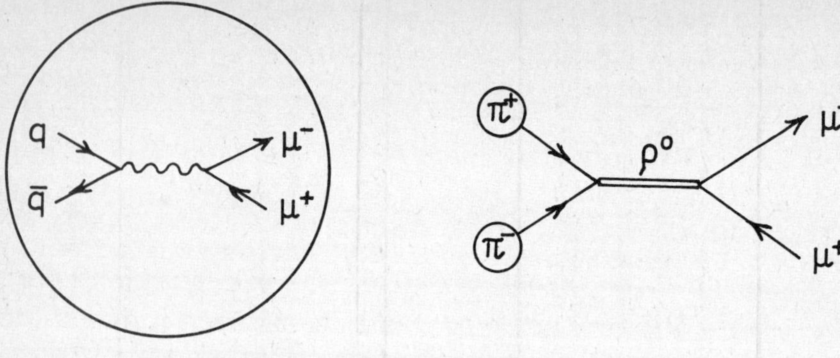

Fig.18: Di-lepton production by quark-antiquark annihilation in the quark-gluon plasma. The main background process is pion annihilation in the ρ channel.

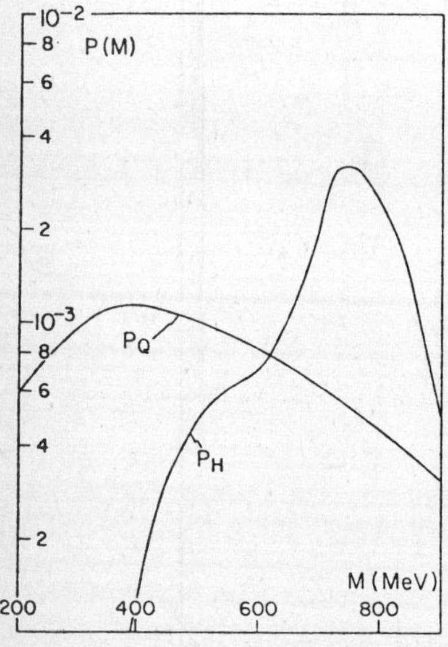

Fig.19: Lepton pair production in the quark (P_Q) and hadron (P_H) phases as fuction of the invariant mass M of the pair (from [DG81]).

While, so far, only production processes involving light (u and d) quarks have been considered, photon or heavy lepton production by strange quarks may also be an efficient process, because strange quarks and antiquarks carry their rest mass in addition to the thermal kinetic energy into the reaction. Photon production by strange quark pair annihilation has recently been studied by Rafelski and Staadt [St85] who find that about one photon from this process can be expected per collision.

(b) Strangeness production:

As the fireball lifetime is much too short for weak interactions to be of importance, strangeness - once produced - can only be destroyed if a strange quark and a strange antiquark meet and annihilate. Since this is not a very likely process unless strange quarks are very abundant, the amount of strangeness observed long after the reaction is over can be expected to be a good signal of the early stages of the evolution of the fireball [Ra81,82]. To find out whether it may act as signature for the quark-gluon plasma, one has to study and compare the abundance of strangeness in the hadronic phase and in the plasma phase.

Let us start with the production of strange hadrons in nucleon-nucleon collisions. As strange antiquarks cannot be contained in a baryon (q=u,d)

$$\Lambda = (uds) \;,\quad \Sigma = (qqs) \;,\quad \Xi = (qss) \;,\quad \Omega = (sss) \;,$$

strange quarks are usually produced in association with a K-meson:

$$p + p \;\rightarrow\; p + \Lambda + K^{+} \qquad , \text{etc.}$$

The threshold energy for this reaction is in the center-of-mass system

$$E^{*}_{cm} = M_{\Lambda} - M_{p} + M_{K} = 570 \text{ MeV} \;,$$

corresponding to slightly more than 1 GeV in the laboratory system.

In contrast, the creation of a strange antibaryon (Λ etc.) is a very extravagant process. It turns out that the easiest way is to produce a pair of Λ-particles

$$p + p \rightarrow p + p + \Lambda + \bar{\Lambda}$$

with a threshold of 2.23 GeV in the center-of-mass system (corresponding to 8 GeV beam energy). Another reaction is

$$p + p \rightarrow p + p + p + \bar{\Lambda} + K^-$$

with a threshold of 2.55 GeV (more than 9 GeV in the lab system). In addition to the high threshold energy these reactions are considerably suppressed because at least three quark-antiquark pairs must be produced in a single reaction and with very similar momenta.

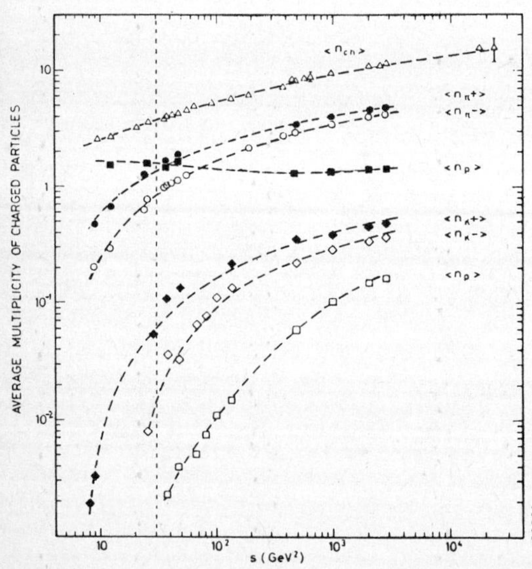

Fig.20: Average particle multiplicities in pp collisions as function of centre-of-mass energy [Ca74].

The difficulty of producing strangeness in hadronic reactions is reflected in the reaction products of, say, a proton-proton collision at

100 GeV beam energy. This corresponds to the energy range (per nucleon) of the proposed nuclear collision experiments at the CERN-SPS [St82]. We find the following table of multiplicities in the final state:

part.	p	n	π^+	π^0	π^-	Λ	K^+	K^-	K^0	$\overline{K^0}$	$\bar{p}$
mult.	1.4	0.5*	2.3	2.0*	1.76	0.11*	0.17	0.10	0.14*	0.10*	0.01

A star "*" indicates that the number has been deduced by simple quark counting arguments. When we break up the final state hadrons into their quark content, we obtain the following table

origin	u	$\bar{u}$	d	$\bar{d}$	s	$\bar{s}$
p	2.8		1.4			
n	0.5		1.0			
π^+	2.3			2.3		
π^0	1.0	1.0	1.0	1.0		
π^-		1.76	1.76			
Λ	0.11		0.11		0.11	
K^+	0.17					0.17
K^0			0.14			0.14
$\overline{K^0}$				0.10	0.10	
K^-		0.10			0.10	
$\bar{p}$		0.02		0.01		
total	6.88	2.88	5.41	3.41	0.31	0.31
prior coll.	4	-	2	-	-	-

To eliminate the effect of isospin asymmetry in the entrance channel we add the contributions of all light quarks, finding

$$\begin{aligned} N_q &= N_u + N_d = 12.3 \\ N_{\bar{q}} &= N_{\bar{u}} + N_{\bar{d}} = 6.3 \\ N_s &= N_{\bar{s}} = 0.31 \; . \end{aligned} \tag{5.1}$$

Of particular interest are the ratios

$$N_s \,/\, N_q \approx 0.025 \quad , \quad N_{\bar{s}} \,/\, N_{\bar{q}} \approx 0.05 \; . \tag{5.2}$$

A more comprehensive analysis of strange particle production in high-energy collisions, which supports our estimate, can be found in the review lecture of M.Faessler [Fa84].

We now compare these results with the abundance of strange quarks in the quark-gluon plasma phase. We shall assume that all quark flavours (u,d,s) take on their thermal and (hadro-)chemical equlibrium distributions. Whether chemical equilibrium is actually reached for strange quarks is a non-trivial question, but calculations to be discussed in section 12 show that this should be the case to a considerable degree. Since strange quarks are always produced in pairs, their chemical potential must be zero. Because they are not sufficiently abundant to form a degenerate Fermi gas, we may use the Boltzmann approximation to the Fermi distribution function:

$$\begin{aligned} N_s = N_{\bar{s}} = 6V \int (dp)\, e^{-\beta\sqrt{(p^2+m_s^2)}} &\approx 3/\pi^2 \, T m_s^2 \, K_2(\beta m_s) \approx \\ &\approx 3/\sqrt{2}\, V (T m_s/\pi)^{3/2} \, e^{-\beta m_s} \;, \end{aligned} \tag{5.3}$$

where V is the fireball volume and the factor 6=(2×3) counts spin and colour degrees of freedom.

With the same approximations we find for the number of light antiquarks contained in the fireball plasma

$$N_{\bar{q}} = N_{\bar{u}} + N_{\bar{d}} = 12V \int(dp)\ e^{-\beta(p+\mu)} \simeq 12/\pi^2\ VT^3\ e^{-\beta\mu} \quad , \tag{5.4}$$

where now the factor 12=(2×2×3) counts flavour, spin and colour degrees of freedom. μ is the light-quark chemical potential, which has a typical value of 300 MeV (1/3 of the nucleon mass). If we put $T \simeq m_s \simeq 200$ MeV, we find for the ratio of strange quarks or antiquarks to light antiquarks:

$$N_{\bar{s}}/N_{\bar{q}} = \tfrac{1}{4}\sqrt{(\pi/2)}\ (\beta m_s)^{3/2}\ e^{\beta(\mu-m_s)} = 1/3\ e^{\beta(\mu-m_s)} \simeq \tfrac{1}{2}\ . \tag{5.5}$$

This is very much different from the ratio deduced from the reaction products of a proton-proton collision, which was 1/20. The quark-gluon plasma contains as many strange quarks and antiquarks as it contains non-strange antiquarks. This different behaviour comes mainly from the fact that hot hadronic matter consists mostly of pions, which are by half made of light antiquarks.

For light quarks in the plasma we find typically

$$N_q = N_u + N_d = 12\ V \int(dp)\ [e^{\beta(p-\mu)}+1]^{-1} \simeq 2/\pi^2\ V\ \mu(\mu^2+(\pi T)^2)\ . \tag{5.6}$$

For the same values of μ and T as above

$$N_s/N_q \simeq 0.04\ , \tag{5.7}$$

which is still a factor of two larger than the same ratio for the p-p collision products.

We conclude that there is a remarkable enhancement of strangeness in the quark-gluon phase over the hadronic phase. Optimal discrimination may be obtained from rare strange antibaryons, such as the ratios $N_{\bar{\Lambda}}/N_{\bar{p}}$, or even

$N_{\bar{\Omega}}/N_{\bar{\Lambda}}$ [Ra84]. An important question in this context is whether time permits the equilibrium distribution of strange quarks to be reached in a high-energy nucleus-nucleus collision. This is a question of reaction rates which will be the subject of ch. 12.

(c) Other signatures:

We briefly mention two other possible signals for the quark-gluon plasma that have recently been suggested. One concerns the production of antiparticles and the other involves charge correlations in rapidity space.

That an enhanced production of antinuclei might be indicative for the formation of a quark-gluon plasma was suggested by Heinz, Subramanian and Greiner [HSG84]. Their argument is based on the observation that an antibaryon must be produced in association with another baryon in the hadronic phase. The threshold for this process is very high (this was already stressed in connection with strangeness, where anti-lambda production is highly suppressed). Although the ratio of antiquarks in the hadronic phase is generally higher than in the quark phase (compare eqs. (3) and (5)), the overwhelming majority of them are contained in mesons, mainly pions, and only very few in antibaryons. On the other hand, in a coalescence model [SY81] there is a fair chance of producing an antibaryon from the antiquarks contained in the quark-gluon plasma during break-up. The ratio is favourably enhanced when antinuclei, containing several antibaryons, are considered.

Heinz et al. predict that antinucleons are more abundant by a factor 3 in the plasma phase, and anti-alpha particles even by a factor 100. The crucial question is whether these enhancements stand a chance of surviving the equilibration processes in the hadronic break-up phase. In view of the large annihilation cross-sections at low energy (100 mb at 1 GeV/c beam momentum, and rapidly rising at lower energy) this seems to be extremely doubtful, at least. A detailed calculation of hadronic reaction kinetics in the break-up phase as well as for a collision proceeding fully in the hadronic phase are required to decide this question (this program is currently being pursued by Koch and Rafelski [KR85]). Also, the possible clustering of quarks or antiquarks in the plasma phase [CCR84] may have a dramatic effect on the predictions for antiparticle cluster formation from the quark-gluon plasma.

Siemens et al. [LPS84] have suggested that the lack of charge correlations between pions of neighbouring momenta may provide a signature for the quark-gluon plasma. In hadronic reactions, quark-antiquark pairs are usually created by the breaking of colour strings attached to the leading partons. At high energy this leads to the formation of particle 'jets' where the quark and the antiquark of a newly created pair end up in different hadrons which are closely related in rapidity (see Fig. 21). This causes an anticorrelation in the charges of nearby hadrons [FF78] which is clearly observed in experiments at DESY [Br81,Al84]. E.g. protons and antiprotons produced in e^+-e^- collisions are ten times more likely contained in the same jet than in opposite jets [Al84].

Fig. 21: Particle jet formation by the breaking of a colour string. Quark and antiquark of the same pair end up in adjacent mesons.

Because of the global equilibration of the colour degrees of freedom in a quark-gluon plasma one would expect that no such anticorrelation arises in the break-up of the quark-gluon plasma into hadrons. Through many scattering processes a quark and its antiquark in the plasma should lose their knowledge of a common origin. Although this picture is intuitively obvious, detailed estimates of the expected charge correlations are missing. In particular, it is not clear whether the anticorrelations observed for collisions of individual elementary particles would not also be washed out by secondary interactions in the dense hadronic matter formed in a nuclear collision that does not produce a quark-gluon plasma.

LECTURE 3

In the third lecture we dicuss our present (theoretical) understanding of the deconfinement phase transition. We start with a brief review of lattice QCD, which lends itself to an evaluation on the computer by the methods of stochastic sampling. Alarge amount of work has been directed toward this approach over the last three years. The main established results are:

(1) there is a long-range force between a quark-antiquark pair in the QCD vacuum, causing confinement of colour,

(2) at finite temperature (of the order of 200 MeV) there is a phase transition to the deconfined phase of the (quark-) gluon plasma.

However, calculations including quarks in a satisfactory way are still in their infancy, so that many questions and most quantitative details are not yet decided.

While the lattice-QCD approach yields rigorous numerical results, these are not easily comprehensible in terms of simple ideas. In the second part of this lecture we, therefore, discuss a much simpler - in fact, oversimplified - model for the true ground state of QCD: the chromomagnetic vacuum. In spite of its simplicity this model describes the energy difference between the true and the perturbative vacuum and the transition to a deconfined, perturbative phase at high temperature, also giving a phase transition temperature of about 200 MeV. We finally estimate the possible suppression of the phase transition due to the existence of a barrier between the two phases.

Fortunately, a numerical non-perturbative approach to finite temperature QCD has become available in the past few years, which allows to treat the interactions exactly. This scheme, known as Monte-Carlo evaluation of lattice QCD is based on a stochastic sampling of the exact functional integral for the partition function of finite temperature QCD defined on a space-time grid.

The partition function for a quantum system described in terms of fields A(x) by a Hamiltonian H[A] is

$$Z = \mathrm{Tr}(e^{-\beta H[A]}) , \tag{6.1}$$

where $1/\beta=T$ is the physical temperature. The lattice formulation is obtained from this general expression in three steps which we shall briefly sketch for the pure SU(3) Yang-Mills theory.

First, the partition function Z is rewritten in the form of a functional integral over Euclidian space-time (i.e. with imaginary time variable $\tau=it$) [Be74] :

$$Z(\beta,V) = \int(DA) \exp[\int d\tau \int d^3x L(A(x,\tau))] , \tag{6.2}$$

where L[A] is the Lagrangian of the Yang-Mills field:

$$L[A] = -1/4 \, (\partial_\mu A_\nu^a - \partial_\nu A_\mu^a + g f_{abc} A_\mu^b A_\nu^c)^2 . \tag{6.3}$$

The integration runs over all field configurations with periodic boundary condition $A(x,\beta)=A(x,0)$. The three-dimensional integral of the Hamiltonian formulation, $H=\int d^3xH(x)$, thus becomes a four-dimensional integral with the extra dimension measuring the temperature.

In the next step, the Euclidian x-τ continuum is replaced by a finite lattice, with N_x sites of spacing a_x in the three spatial directions, and N_τ sites of spacing a_τ in the temperature direction. The integrals in the exponent of eq.(2) are thus converted into finite sums, and we have $V=(N_x a_x)^3$, $\beta=N_\tau a_\tau$. The continuum theory is recovered in the limit $a_x, a_\tau \to 0$ with the product $N_\tau a_\tau$ being kept fixed, while the thermodynamic limit requires $N_x \to \infty$ at fixed spacing a_x.

In the last step, following Wilson [Wi74], we replace the gauge field variable $A[\frac{1}{2}(x_i+x_j)]$ associated with the link between two adjacent lattice sites i and j by the gauge transformation

$$U_{ij} = \exp[-i\tfrac{1}{2}\lambda_a (x_i-x_j)^\mu A_\mu^a(\tfrac{1}{2}(x_i+x_j))]. \tag{6.4}$$

U_{ij} is just the rotation in colour space required to compare the colour coordinate systems at the two adjacent lattice sites as we discussed in chapter 1. In this way the integration over all field configurations is transformed into an integration over all elements of colour SU(3) at each lattice link:

$$Z(\beta,V) = \int\Pi_{links}(dU_{ij}) \exp[-S(U)] \tag{6.5}$$

where Wilson's lattice action for SU(3) is given by a sum over all (elementary) plaquettes ijkl

$$S(U) = 6/g^2 \, (a_\tau/a_x)^{\pm 1} \, \Sigma_{(ijkl)} \, (1 - 1/3 \text{ Re Tr } U_{ij}U_{jk}U_{kl}U_{li}) \, . \tag{6.6}$$

The plus sign holds if all links of the plaquette are space-like, the minus sign applies if two links run in the imaginary-time direction. The continuum theory is recovered as leading term for $a \to 0$ by expanding the exponential (4). Expressions for the energy density, etc., are easily derived from eqs.(5,6). All these expressions involve an average over all plaquettes, i.e. elementary squares, on the lattice.

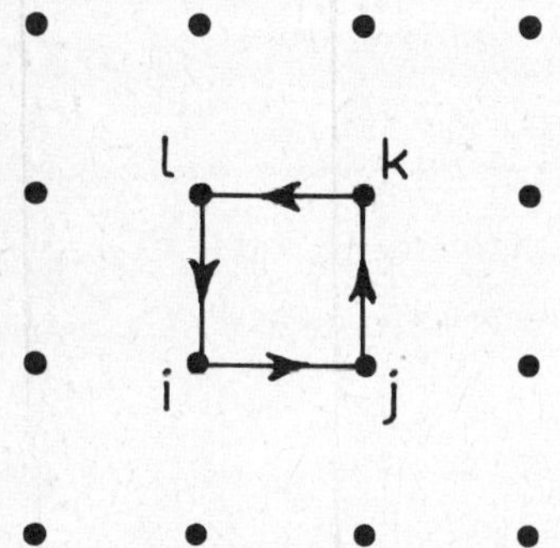

Fig.22: Definition of the SU(3) action on the lattice as sum over plaquettes.

This is the starting point for the Monte-Carlo evaluation of gluon thermodynamics [Cr80]. The computer initially assigns a group element U_{ij} to each lattice link (ij). Typically one chooses either U=1 for all links (this is called a 'cold start') or one chooses U_{ij} randomly ('hot start'). Then every link is successively reassigned with a new element chosen randomly with probability weight exp[-S(U)]. One full sweep through the entire lattice is called one iteration. The weight clearly favours configurations with a small action, i.e. the lattice tends on the average towards the configuration with least action. However, the stochastic choice of new link variables guarantees that the scheme samples all possible configurations, if only it is run long enough. In general, it is found that a few hundred iterations provide first indications about the behaviour of the energy density, but for some precision the sample should contain many more iterations. Unfortunately, for SU(3) present computer limitations have not yet allowed to generate more than the minimal sample. The success of the approach rests on the empirical fact that already rather small lattices (N_x=5-10, N_τ=3-5) seem to yield asymptotic answers with respect to the thermodynamic limit as well as to the continuum limit. One reason for this finding may be the very large number of links, i.e. of independent variables, contained even in a 'small' lattice (there are $4N_x{}^3N_\tau$ links, i.e. about 12000 for a 10×10×10×3 lattice!).

As a result of the Monte-Carlo calculation, one obtains the energy density $\varepsilon=E/V$ as function of the coupling constant g. In the continuum limit, g and the lattice spacing a are related through the renormalization group equation yielding

$$a\Lambda_L = (16\pi^2/11g^2)^{51/121}\, e^{-8\pi^2/11g^2} . \tag{6.7}$$

This equation is obtained by requiring that the dimensional parameter Λ_L be invariant under scale changes. All results are then expressed in terms of the lattice parameter Λ_L. In oder to obtain results in physical units, Λ_L must be adjusted by 'measuring' one physical quantity, e.g. by comparing with the value of the string constant as derived from the slope of the Regge trajectories. Typical values of Λ_L lie in the region $\Lambda_L \simeq 2$ MeV.

To see where the confinement phase transition occurs, it is useful to plot the ratio of the energy density ε to the value ε_{SB} obtained from the Stefan-Boltzmann law for an ideal gluon gas. As discussed in chapter 9, this ratio $\varepsilon/\varepsilon_{SB} = D_{eff}$ describes the reduction of the effective number of degrees of freedom due to the presence of interactions. Evaluated with the Monte-Carlo technique on the lattice, this number is in principle exact and is not based on model assumptions. For SU(3) one has found [MP82,CES83] that $\varepsilon/\varepsilon_{SB}$ suddenly drops from one to zero when the temperature T falls below about 80 Λ_L, corresponding to a physical value of the critical temperature T_c = 208±20 MeV. More detailed investigations [En81a] have revealed that the system behaves essentially like a gas of glueball states, i.e. that single gluons are missing from the excitation spectrum, at temperatures below T_c.

More recently, extended calculations have revealed that the continuum limit is not reached as easily on the lattice as was initially thought

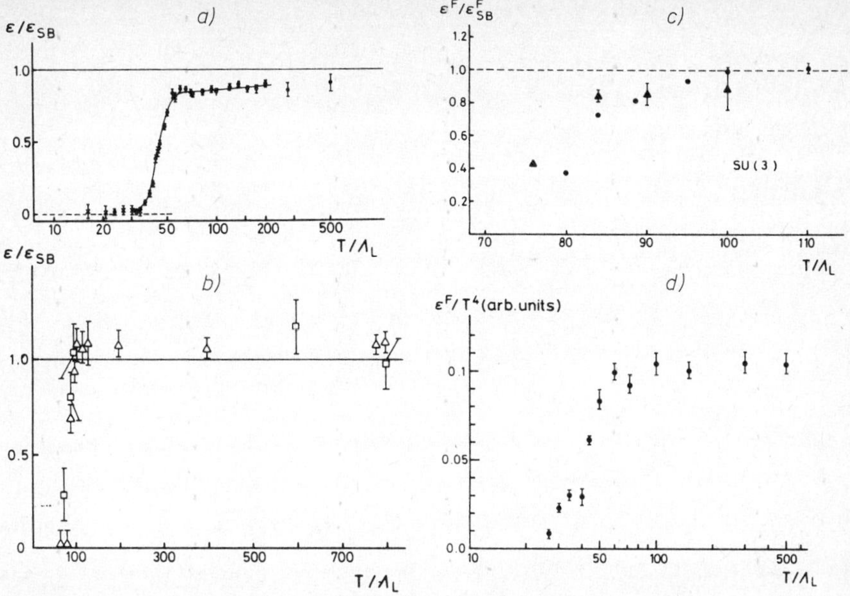

Fig.23: Ratio $\varepsilon/\varepsilon_{SB}$ (true energy density relative to that of a gas of non-interacting massless particles) as obtained by Monte-Carlo evaluation on the lattice. (a) Pure SU(2) gauge theory [En81,82], (b) pure SU(3) gauge theory [MP82], (c) SU(3) with quarks [EKS82], (d) SU(2) with quarks [Sa82].

[Ke84]. One has to go to fairly large lattices to even determine the critical temperature with some precision. The present calculations are not yet accurate enough to allow for a precise determination of more complex physical quantities such as the latent heat in the phase transition.

The extension of lattice QCD to include quarks is much less developed. There are several reasons for this. For one, there does not seem to be such a natural way to introduce fermions on the lattice as the one employed for the gauge field itself, where one simply defined gauge group elements on the links. Fermions must be defined on the lattice sites, and the derivative $\partial_\mu\psi$ must be approximated by finite differences. It is not surprising that this comparatively clumsy scheme, which is not even unique, leads into problems like spectrum doubling, etc. [Su77]. The second difficulty originates in the Pauli principle which requires that fermion fields anticommute. The functional integral over $\psi(x)$ becomes a determinant running over the lattice sites. One is thus confronted with the problem of calculating determinants of huge size, for which a satisfactory solution has yet to be found. Several methods have been devised to approximately treat the effect of real quarks, but virtual quark-antiquark pairs have been included in the calculations only very recently.

When virtual quark pairs are included, i.e. in real QCD, there exists a further order parameter besides the Wilson loop which controls colour confinement. This order parameter is the vacuum expectation value of the scalar quark density, $\langle\bar{\psi}\psi\rangle$. When it is zero, the vacuum is invariant under chiral transformations. When it is non-zero, $\langle\bar{\psi}\psi\rangle$ acts as an effective quark mass term and the vacuum breaks chiral symmetry. This is the case in QCD at low temperatures, but a phase transition to a chirally invariant phase is expected at or above the deconfinement phase transition [PW84]. Also doubts exist whether deconfinement occurs as a true (discontinuous) phase transition in the presence of virtual quark pairs [BU83].

During the past year, several groups have presented first results of Monte-Carlo simulations including virtual quark-antiquark pairs [Po84,FRS84,GLP84]. The general picture arising from these numerical data is the following: At low temperatures QCD is in a chirally broken phase with colour confinement. At about the same temperature (of the order of 200 MeV) the expectation value of $\bar{\psi}\psi$ goes to zero and that of the Wilson loop begins to rise. In parallel, the energy density increases rapidly and approaches the density of a gas of free quarks and gluons, with a possible region of overshooting right after the phase change (see Fig. 24).

What are the conclusions to be drawn from the results of these formidable numerical calculations? The phase transition predicted for QCD, indeed,

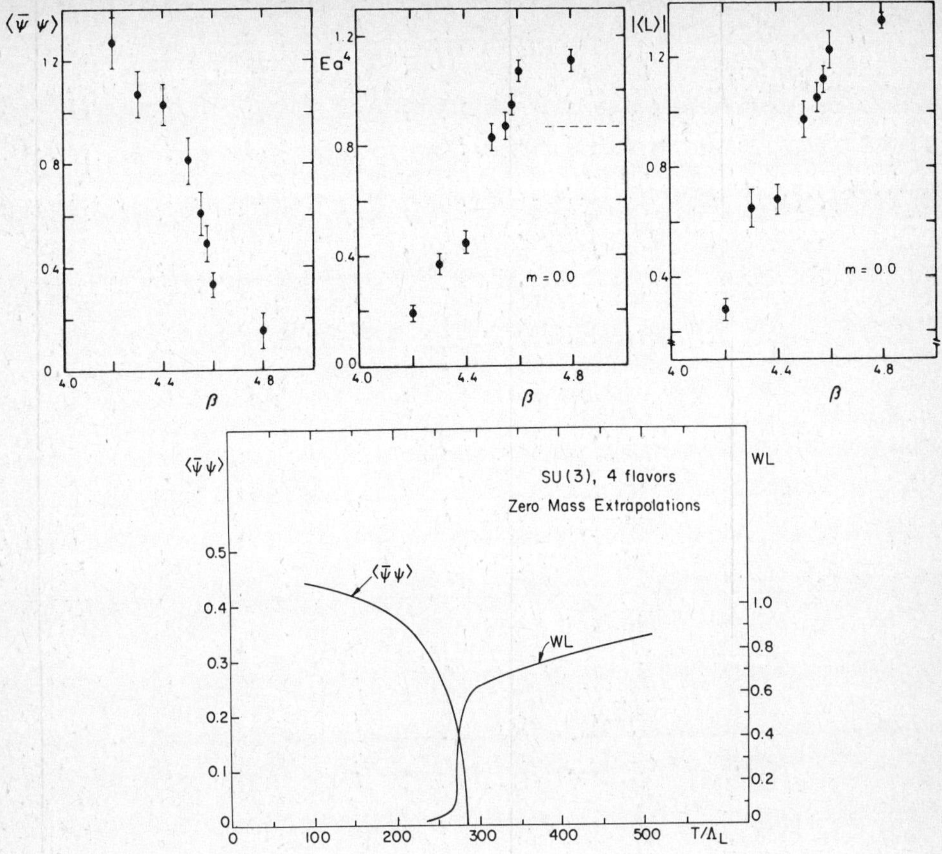

Fig. 24: Chiral and deconfinement phase transition in lattice QCD with virtual quarks. Upper part: Quark mass term, energy density and Wilson loop (from [GLP84]), lower part: mass term and Wilson loop (from [Po84]).

seems to be in the temperature range between 150 and 200 MeV as expected from simple arguments. The phase transition leads from a low temperature phase of colour confinement into the coloured plasma phase [MS81,KPS81]. For the pure SU(3) gauge theory, the phase transition is of first order with a substantial latent heat caused by the rapid thawing of the colour degrees of freedom.[1] The quantitative value of this latent heat is not yet firmly established, results range from $\Delta\varepsilon \simeq 3.75\ T_c^4 \simeq 0.9$ GeV/fm^3 [CES83] to $\Delta\varepsilon \simeq 1.9$ GeV/fm^3 [SF83]. For the full theory including quarks there is also a chiral phase transition approximately at the same temperature. Quantitative details of this picture are not yet firmly established, but must be determined by future improved calculations.

---

[1] The order of the phase transition seems to be determined by details in the theoretical model. E.g., the pure SU(2) gauge theory is found to have a second order phase transition to the deconfined phase, in contrast to the first order transition found in SU(3).

To implement the confinement of quarks and gluons in the bag model we have postulated that the true ground state of QDC has a complicated structure. We know some of the properties it must have:

- its energy density must be lower than that of the perturbative vacuum (this difference is related to the bag constant),
- it must be a perfect colour dielectric, i.e. $\varepsilon=0$, $\mu=\infty$ (to guarantee confinement),
- it must be Lorentz and gauge invariant (the vacuum has to be!).

Since we expect a phase transition into the perturbative vacuum, there must be an order parameter describing this phase transition. The order parameter must vanish in the perturbative phase and must be non-zero for the true ground state. What may this order parameter be? Because of the symmetry requirements it must be a gauge-invariant Lorentz scalar. For the gauge field alone, it can only be composed of scalar combinations of the field strength tensor, such as

$$\langle F^a_{\mu\nu}(x)F^{\mu\nu}_a(x)\rangle \quad , \quad \langle d_{abc}F^a_{\mu\nu}(x)F^{\nu\lambda}_b(x)F^{a\mu}_\lambda(x)\rangle \ , \tag{7.1}$$

or, if we also permit non-local expressions involving a structure function S(x-y):

$$\langle \int d^4y F^a_{\mu\nu}(x)S(x-y)F^{\mu\nu}_a(y)\rangle \ , \text{ etc.} \tag{7.2}$$

For full QCD also bilinear expressions involving quark fields may occur as order parameter, e.g.

$$\langle\bar{\psi}(x)\psi(x)\rangle \ . \tag{7.3}$$

Although it is, in principle, possible to investigate the energy of a trial ground state as a function of these order parameters, this would be a very hard problem to solve. However, nuclear physicists know that often a well-chosen trial state may yield a good estimate of the ground state energy and static properties of a nucleus, even if it does not take care of all the required symmetries. Just think of the success of the deformed shell model, the Nilsson model, in explaining nuclear deformation parameters, magnetic moments, etc.! Although the restoration of the artificially broken symmetry gives some, often small, correction to the ground state energy, its real importance shows up only in transition matrix elements between different states.

In this spirit, one is led to consider the expectation value $\langle F^a_{\mu\nu}(x)\rangle$ itself as an order parameter. To be sure, the state described by a non-zero expectation value of $F^a_{\mu\nu}$ cannot be the true ground state, because it violates the invariance properties of the vacuum state, but its energy expectation value may nonetheless be a good approximation to the ground state energy [PT78].

For simplicity, let us start by discussing SU(2) gauge theory instead of full SU(3). A constant colour-electric field is not a reasonable choice for the approximate ground state, because it accelerates coloured particles and leads to immediate pair-creation. Therefore, let us study the energy of the gluon vacuum in the presence of a constant chromomagnetic field of strength H, which we may take, without loss of generality, to point in the colour-3 direction and into z direction in coordinate space:

$$\vec{H}_a = H\vec{e}_z\delta_{a3} \ , \quad \vec{A}_a = Hx\vec{e}_y\delta_{a3} \ . \tag{7.4}$$

We shall now investigate the properties of small fluctuations around this field configuration. The non-linear terms in the Yang-Mills equations couple the three different colour components among each other through the SU(2) structure constants ε_{abc}, where (abc) must be a permutation of (123). As a consequence, the fluctuations in the colour-3 direction do not feel the presence of the average chromomagnetic field (in lowest order), while the fluctuations in the a=1,2 directions will be influenced by the background field.

It is convenient to express the two modes by a single complex field $W_\mu = (A^1_\mu + A^2_\mu)/\sqrt{2}$. This corresponds to a transition from cartesian to spherical coordinates in colour space. Keeping only terms of second order in the field W, the part of the total SU(2) Lagrangean governing small oscillations in the colour-1,2 directions is [NO78]

$$L_2(W) = -\tfrac{1}{2}|(\partial_\mu - igA^3_\mu)W_\nu - (\partial_\nu - igA^3_\nu)W_\mu|^2 - ig(\partial_\mu A^3_\nu - \partial_\nu A^3_\mu)W^*_\mu W_\nu \,, \tag{7.5}$$

from which the equations for the W-modes may be obtained by variation with respect to W^*_μ. The gauge freedom can be utilized to simplify these equations by imposing the so-called background field gauge

$$(\partial_\nu - igA^3_\nu)W^\nu = 0 \,. \tag{7.6}$$

Furthermore, the expression involving A^3_μ in the last term in eq. (5) can be identified with the field strength tensor $F^3_{\mu\nu}$ of the background field, because the nonlinear term vanishes for the special field configuration (4). We thus find the linearized equation of motion:

$$[(\partial_\lambda - igA^3_\lambda)^2\delta_{\mu\nu} + 2igF^3_{\mu\nu}]\, W^\nu(x) = 0 \tag{7.7}$$

for small amplitude oscillations of the W-field.

Exactly the same equation would be obtained in quantum electrodynamics of a charged vector (spin 1) field in the presence of an electromagnetic background field. (The mass of the particle is zero, which makes the comparison academic since there is no known electrically charged massless particle in nature.) The first contribution describes motion due the interaction of the charge of the particle with the field, while the second term contains the effect of the field on the magnetic moment of the spin-1 particle. The factor 2 is nothing else than the g-factor for a point-like particle. This can be seen as follows: restricting to space-like values of the vector indices $\mu,\nu=1,2,3$, the matrix $iF^3_{\mu\nu}$ can be identified with the spin-matrix for a spin-1 particle (the three-dimensional representation of the rotation group):

$$iF^3_{\mu\nu} = H \begin{pmatrix} 0 & i & 0 \\ -i & 0 & 0 \\ 0 & 0 & 0 \end{pmatrix} = -HS_z \ . \tag{7.8}$$

For the space-like components of W_ν the second term in eq. (7) can therefore be written in the suggestive form

$$(2g\vec{H}\cdot\vec{S})\ \vec{W} \tag{7.9}$$

the interpretation of which is obvious.

For stationary states

$$\vec{W}(x) = \vec{\xi}_s \exp(ik_2y+ik_3z-i\varepsilon t)\ W_{s,k_2k_3}(x) \tag{7.10}$$

the eq. (7) reduces to the equation for the Landau orbits of a charged particle in a magnetic field

$$[-\varepsilon^2-\partial^2/\partial x^2-(k_2-gHx)^2+k_3{}^2-2gHs]\ W_{s,k_2k_3}(x) = 0 \ . \tag{7.11}$$

Here $\vec{\xi}_s$ (s=-1,0,1) are the eigenvectors of the spin matrix S_z. Because gluons are massless, the eigenvector with s=0 is ruled out, i.e. only the two transverse polarizations $s=\pm1$ exist. The energy eigenvalues determined by eq. (11) are well-known:

$$\varepsilon_{ns}(k_3) = [gH(2n+1) + k_3^2 - 2gHs]^{\frac{1}{2}} , \tag{7.12}$$

where $n=0,1,2,...$, $s=\pm1$ and $-\infty<k_3<\infty$. Due to the absence of a mass term in eq. (12), the radicand becomes negative for the mode $n=0$, $s=+1$ when $k_3^2<gH$. The interaction between the spin of the gluon and the chromomagnetic field is so strong that it overwhelms the zero-point energy of Landau motion. Mathematically speaking, the occurrence of imaginary energies means that the operator in eq. (7) is not self-adjoint. In physical terms this means that one is not dealing with a well-posed problem.

The reason for the unwanted behaviour is the homogeneity of the chromomagnetic field. If the field would, e.g., oscillate in the transverse direction, then the interaction with the spin of gluons in the lowest mode would at least partially cancel, because $2g\langle H\rangle s$ would involve the average of H over the entire gluon mode. It is clear that the effective spin interaction could be made arbitrarily small for sufficiently fast varying orientation of the background field. Now imagine the background field to become more and more homogeneous, starting out from a rapidly oscillating field configuration. The spin interaction would gradually increase, causing the energy eigenvalue of the lowest Landau mode to decrease until it would finally reach zero. At that moment, gluons would spontaneously be produced from the vacuum forming a condensate in the supercritical lowest mode which, being spin-oriented, would provide an extra contribution to the average chromomagnetic field. Since the lowest mode is localized in a region of size $(gH)^{\frac{1}{2}}$, the additional field from the condensate has structure of that scale. We conclude that also the average chromomagnetic field must exhibit structure on the scale of the size of the lowest Landau orbit. Any attempt of making it more homogeneous would result in the reappearance of the structure via spontaneous gluon creation. Observe incidentally, that the mathematics is cured at the same time, because the linear approximation does not work for a condensate, so that the full nonlinear eigenvalue problem must be studied.

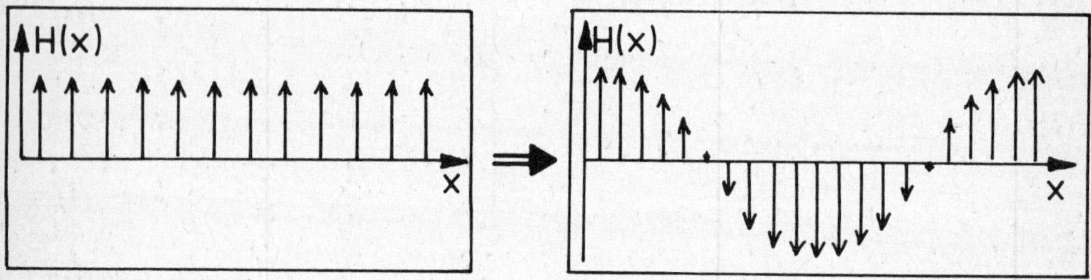

Fig.25: A homogeneous chromomagnetic field is unstable against local fluctuations in the orientation of the field. A gluon condensate develops, causing a domain structure of the field configuration.

This nonlinear problem has not yet been solved, although some attempts have been made in this direction [NO78a,NN79,NO79]. To keep our model as simple as possible - remember that our state is not expected to have good symmetries, anyway - we simply discard the worrisome modes ($n=0$, $s=1$, $k_3^2<gH$). We shall denote this "kitchen recipe" by a prime on the summation sign. The vacuum energy is then just the sum of the zero-point energies of all modes, where we subtract the energy of the perturbative vacuum (i.e. for $H=0$) as the reference point:

$$\begin{aligned} E_V(H) - E_V(0) &= \tfrac{1}{2}\sum_i{}' \varepsilon_i(H) - \tfrac{1}{2}\sum_i \varepsilon_i(0) = \\ &= \tfrac{1}{2}V \, \Sigma_C \Sigma_s \left[gH/\pi \sum_n{}' \int dk_3/2\pi \; \varepsilon_{ns}(k_3,H) - \int d^3k/(2\pi)^3 |k| \right]. \end{aligned} \tag{7.13}$$

The additional sum Σ_C runs over the degrees of freedom of the gluon field in colour space. For SU(3) this will contribute a factor 3. To get rid of the square roots, the trick is to introduce the integral representation

$$\varepsilon^{\alpha} = \Gamma(-\alpha)^{-1} \int_0^{\infty} ds/s \; s^{-\alpha} \; e^{-s\varepsilon^2} \tag{7.14}$$

for $\alpha=\frac{1}{2}$. The sums over n and s yield geometric series, while the momentum integrations reduce to Gaussian integrals (some care must be taken for the lowest mode). Substituting $x=gHs$, the final integration becomes

$$E_V(H) - E_V(0) = -3V(gH)^2/8\pi^2 \int_0^{\infty} dx/x^2 \; [1/\sin(x) - 1/x - e^{-x} + e^{x}\Gamma(\tfrac{1}{2},x)/\Gamma(\tfrac{1}{2})] \tag{7.15}$$

It turns out that the integral is divergent due to the $1/x$ behaviour of the integrand in the limit $x\to 0$. The coefficient of this term is found to be $-1/6 + 2 = 11/6$. The divergent part is easily separated by dimensional regularization, replacing $\alpha=\frac{1}{2}$ by $\alpha=\frac{1}{2}-\varepsilon$ in eq. (14). The integral then splits into a finite part (which is uninteresting) and an infinite part proportional to $(gH)^2$:

$$\begin{aligned} \int dx/x^{1-\varepsilon} (gH)^{2-\varepsilon} &= (gH)^{2-\varepsilon}/\varepsilon = (gH)^2[1-\varepsilon \ln(gH)]/\varepsilon = \\ &= (gH)^2/\varepsilon - (gH)^2 \ln(gH) \, . \end{aligned} \tag{7.16}$$

Since QCD is a renormalizable theory, all terms that go like $(gH)^2$ serve to renormalize the coupling constant in the free Lagrangean $\frac{1}{2}H^2$, and only the finite logarithmic contribution remains:

$$E_V'(H) = [E_V(H) - E_V(0)]_{ren} = 3V/8\pi^2 \; 11/12 \; (gH)^2 \ln(H^2/H_0{}^2) \, , \tag{7.17}$$

where H_0 is a constant depending on the renormalization point. Since H_0 is introduced by the renormalization procedure, its value cannot be calculated but must be determined by comparison with measured quantities.

Despite its simplicity, which corresponds with the simplicity of our choice of the trial vacuum state, this result (first obtained by Savvidy et al. [BMS77,Sa77]) has interesting properties.[2] Because the logarithm is negative for small values of the chromomagnetic field H, the energy of our trial state has a nontrivial minimum given by

$$\partial E_v'(H)/\partial H = 0 \quad \rightarrow \quad H_{min} = H_0 \exp(-\tfrac{1}{2}) \ ,$$

where

$$\Delta\varepsilon = E_v'(H_{min})/V = -11/32\pi^2 \ (gH_{min})^2 = -11/32\pi^2 e \ (gH_0)^2 \ . \tag{7.18}$$

The full curve $\varepsilon_{vac}'(H)$ is shown in Fig.26 where the existence of the nontrivial minimum is obvious. We conclude that the perturbative QCD vacuum is unstable against spontaneous formation of a chromomagnetic field H. The "true" vacuum state is characterized - in this model - by a nonvanishing expectation value of H, and is lower in energy by an amount $\Delta\varepsilon$ given in eq. (18). If we are very courageous we may identify this energy difference with the bag constant B of the MIT bag model, obtaining

$$\varepsilon_v'(H) \ = \ 11/32\pi^2 \ (gH)^2[\ell n(gH)^2/B \ - \ \ell n(32\pi^2/11)-1] \ . \tag{7.19}$$

However, we should bear in mind that in deriving this result we have simply ignored the unstable gluon modes. The true ground state must have even lower energy. Many authors have discussed details of this simple model for the QCD ground state, e.g. the question of confinement, or have

---

[2] The sign of $E_v'(H)$ is opposite to that of the Euler-Heisenberg effective action in QED [Sch58]. This is a consequence of the asymptotic freedom of QCD and, indeed, the factor 11/12 in eq.(17) is intimately related to the factor 11 in the running coupling constant, eq.(1,21) [BMS77]. The "wrong" sign was noticed by Vanyashin and Terent'ev [VT65] as early as 1965, who did not draw the conclusion as to asymptotic freedom, as the Yang-Mills theories were not yet known to be renormalizable at that time.

tried to improve on the trial state. The interested reader is referred to the literature [NO78a,NN79,NO79,AO79,Fl80,Fu80,Ad81].

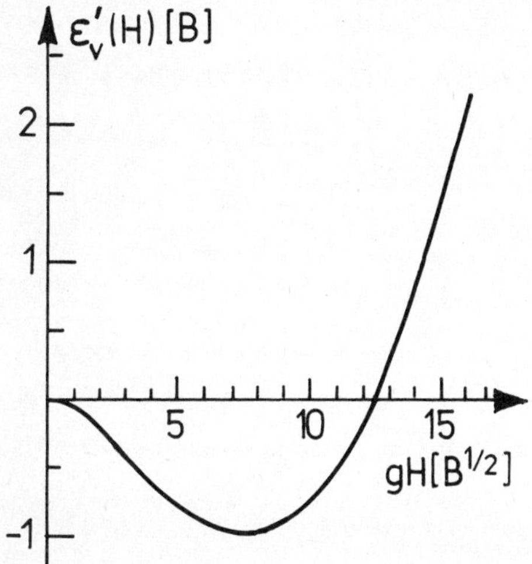

Fig.26: Energy density of the chromomagnetic trial ground state as function of the magnetic field strength. The minimum at a non-zero value of H corresponds to the non-trivial ground state of the gauge theory. The difference in the energy density to the state with H=0 is identified with the bag constant B.

Here, in the context of our survey of the quark-gluon plasma, we will only pursue the question how the colour-magnetic vacuum responds to thermal excitation [MR81,Ka81]. Do we find a phase transition to the perturbative phase, characterized by H=0, beyond a certain critical temperature? The answer is: Yes! To show this, one must add the energy in the thermally excited gluon modes to that of the zero-point fluctuations. The sum in eq.(13) is then modified accordingly to represent the free energy of the gluon field in the presence of an average colour-magnetic field:

$$F(H,T) = \Sigma' \varepsilon_i(H)[\tfrac{1}{2}+(e^{\varepsilon_i(H)/T}-1)^{-1}] = \tfrac{1}{2}\Sigma' \varepsilon_i(H)\mathrm{cth}^{-1}[\tfrac{1}{2}\varepsilon_i(H)/T] \tag{7.20}$$

After some algebra one finds that this is equivalent to the introduction of an additional factor

$$1 + 2\,\Sigma_{\nu=1}^{\infty} \exp(-\nu^2 gH/4xT^2) \tag{7.21}$$

in the integrand of eq.(15), where the "1" stands for the vacuum contribution and the sum contains the thermal excitations. The thermal part is finite and can be easily evaluated on a computer[3].

The result is shown in Fig.27, where the free energy $F'(H,T)=F(H,T)-F(0,T)$ is plotted normalized to its value (3,2), $-\varepsilon_g N_g$, in the perturbative phase. For T=0 the "old" curve $E_v'(H)$ is reproduced, but for increasing temperature the minimum at $H=H_{min}$ is found to become shallower. At a critical temperature T_c, which has been estimated at $1.4B^{\frac{1}{4}}\simeq 200$MeV, the nontrivial minimum and the perturbative phase have the same energy. For $T>T_c$, the phase with H=0 has the lowest free energy. Initially the minimum at $H\neq 0$ still persists as a metastable phase, but the barrier vanishes rapidly with increasing temperature. The existence of a barrier between the two phases at the phase transition temperature T_c is a clear sign of a first order phase transition. This is in agreement with the best available data on SU(3) Monte-Carlo calculations.

The existence of a metastable phase in the vicinity of the phase transition point indicates that superheating (of the confined phase) and supercooling (of the plasma phase) are possible. This possibility may have spectacular observable consequences. For example, as discussed by Gyulassy et al. [Gy84] and by van Hove [vH83], the transition from a supercooled quark-gluon plasma into the confined hadronic phase could proceed in the form of a violent implosion. This might result in anomalously large transverse momenta of the emitted hadrons. Indeed, obvious signs of clustering in rapidity space have been noticed in ultra-high energy nucle-

---

[3] It is important to leave the unstable modes out of the sum in eq.(20) because their instability contradicts the assumption of a thermal equilibrium [CKS83]. Several authors, who have not taken this measure of precaution, have obtained unreasonable results [DS81,NS81].

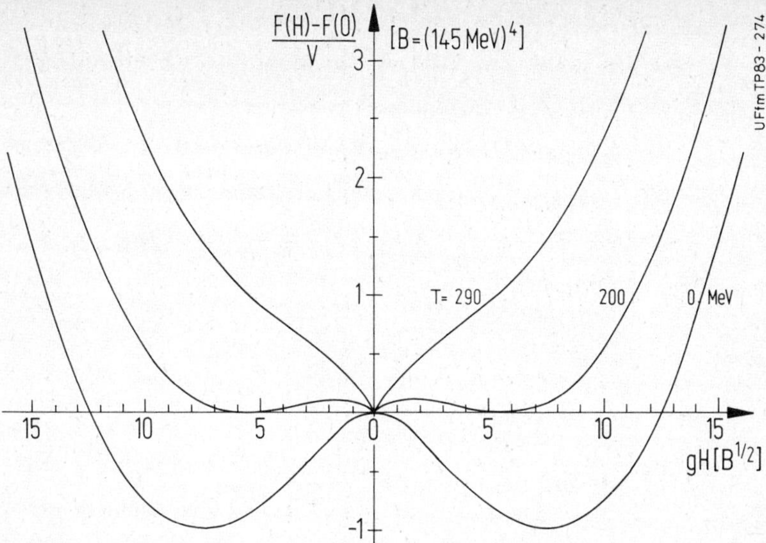

Fig.27: Free energy of the chromomagnetic vacuum trial state as function of field strength H for various temperatures. F is normalized to its value for H=0. At $T \simeq 1.4B^{\frac{1}{4}}$ corresponding to ≃ 200 MeV there is a first order phase transition from the minimum at $H \neq 0$ to the perturbative state at H=0.

ar collisions of cosmic-ray particles by the JACEE collaboration [Bu83,84].

In our model, the possible life-time of a supercooled (or superheated) phase is determined by the time required to decay through the barrier separating the two phases[4]. Let us estimate this time. We shall do so by using the average chromomagnetic field as a collective variable describing the state of the system. The plasma phase is characterized by H=0, the confined phase by the value of H at the nontrivial minimum, H_{min}. The

---

[4] This is only true in the absence of inhomogeneities that my act as nucleation centres for the phase transition. Nucleation will in general reduce the life-time of the metastable state.

typical tunneling time is determined by two counteracting mechanisms: On the one hand, the system would like to spend as little time as possible in the classically forbidden region under the barrier. On the other hand, any change in H is connected with a chromoelectric field $\nabla\times E = -\partial H/\partial t$ that contributes to the energy and increases the effective height of the barrier. The most probable tunneling path finds a compromise between these two obstacles.

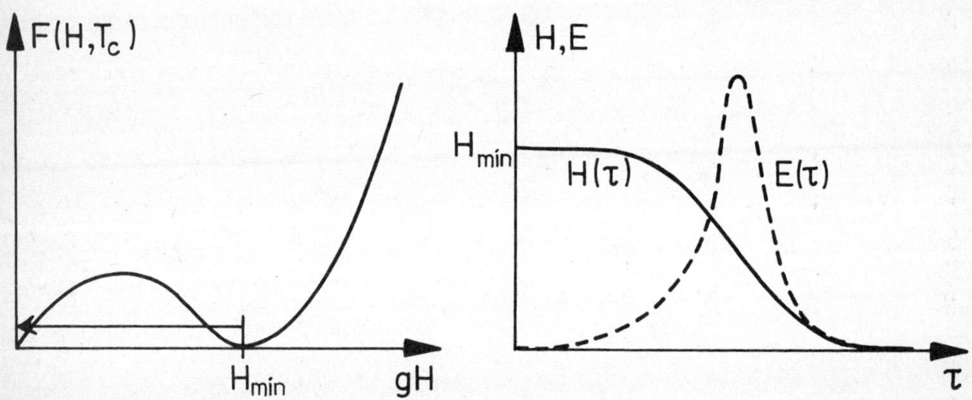

Fig.28: Phase change by tunneling through the barrier in the mean field H. During the period when H varies a chromoelectric field E builds up which is proportional to $\partial H/\partial t$ and adds to the height of the barrier.

For simplicity, let us assume that the transition from the metastable phase to the stable phase occurs in a finite volume of linear extension R. Let us, further, restrict our considerations to global variations of H in that volume. Under these restrictions, the tunneling process is described by a potential function

$$\vec{A}(x,t) = \tfrac{1}{2}\vec{H}(t)\times\vec{r}\ , \qquad |\vec{H}(t)| = \lambda(t)H_{min} \tag{7.22}$$

with $\lambda(t_i)=\lambda_i$, $\lambda(t_f)=\lambda_f$ taking the values 0 or 1. For the transition from the (superheated) confined phase to the plasma phase we have $\lambda_i=1$, $\lambda_f=0$, and vice versa for the supercooled transition.

The total action connected with the tunneling process is then

$$\begin{aligned} S[\lambda(t)] &= \int dt \int d^3x [\tfrac{1}{2}E(t)^2 - F(H(t))] = \\ &= \int dt \int d^3x [(\partial\lambda/\partial t)^2 (\vec{H}_{min} \times \vec{r})^2/8 - F(H_{min}\lambda(t),T)] . \end{aligned} \tag{7.23}$$

The optimal function $\lambda(t)$ which describes the best tunneling path is determined by the stationary point $\delta S/\delta\lambda(t)=0$ [BC78]. Because we are dealing with a tunneling process it turns out that the resulting differential equation for $\lambda(t)$ can only be solved for imaginary time $t=i\tau$, and the action S itself is imaginary for the optimal path. Calling the coefficient of $(\partial\lambda/\partial t)^2$ a generalized collective mass parameter

$$M = \int d^3x \ \tfrac{1}{4}(\vec{H}_{min} \times \vec{r})^2 = H_{min}{}^2 R^5/6 , \tag{7.24}$$

the penetration factor for tunneling through the barrier is given by the generalized Gamov formula

$$P = \exp(-\mathrm{Im}\, S) = \exp[-\int_{\lambda_i}^{\lambda_f} d\lambda \ \{2M \int d^3x (F(\lambda H_{min}) - F(\lambda_i H_{min}))\}^{\frac{1}{2}}] . \tag{7.25}$$

At T_c we have $F(H_{min})=F(0)$ and, therefore, with the volume $\int d^3x = R^3$ we obtain

$$\begin{aligned} P(T_c) &= \exp[-H_{min} R^4 \int_0^1 d\lambda/\sqrt{3} \ \{F(H_{min}\lambda) - F(0)\}^{\frac{1}{2}}] \simeq \\ &\simeq \exp(-\tfrac{1}{2} g H_{min}{}^2 R^4) \simeq \exp(-20\ BR^4) . \end{aligned} \tag{7.26}$$

This result implies that the barrier forms a serious obstacle[5] for tunneling if the phase transition occurs in a region of size larger than $R = \frac{1}{2}B^{-\frac{1}{4}} \simeq (gH_{min})^{\frac{1}{2}}$.

If, for some reason, the phase transition cannot occur independently in smaller regions inside the quark-gluon plasma, the suppression due to the barrier may cause drastic supercooling of the plasma phase. One scenario which comes to mind immediately is that there exist long-range colour correlations in the plasma. A larger region inside the plasma fireball which is not altogether a colour singlet cannot make the phase transition to the confining phase as a whole, since colour must be conserved in the transition process. Under such circumstances supercooling of the plasma phase could occur during the expansion of the hot "fireball" filled with quark-gluon plasma, if the transition is of first order.

---

5 Whether the transition proceeds easily in regions of smaller size, depends on surface effects. This is so, because for dimensions below $(gH)^{\frac{1}{2}}$ the neglect of surface effects due to the phase jump at the boundary of the volume is not warranted. Such surface contributions are known to result in a minimal size of phase transition regions in solid state physics. Lacking a detailed calculation including gradients of the chromomagnetic field H we can anticipate that in our model $R \simeq (gH_{min})^{\frac{1}{2}}$ will play the role of a critical length for the transition volume ("critical droplet").

LECTURE 4

The fourth lecture is devoted to the interactions in the quark-gluon plasma. We first approach the problem from the point of view of perturbation theory, calculating the effective coupling constant $\alpha_s(q,T)$ at finite temperature for static interactions. The result shows that colour interactions are screened at distances larger than 1-1.5 fm due to the high density of coloured particles in the plasma (Debye screening). We also find that the effective coupling strength increases dramatically upon approach of the phase transition temperature, estimated here at $T_c \simeq 170$ MeV. However, contributions of higher order in α_s are seen to be very large, throwing serious doubt on the credibility of the results of perturbative calculations.

In the second part we investigate a special non-perturbative effect of the interactions. Due to colour conservation the whole quark-gluon plasma fireball must form a colour singlet. This global colour symmetry is found to suppress the effective number of degrees of freedom that are active in the plasma. The extent of this suppression depends on the total volume, being especially severe for small regions. Other finite size corrections are also considered, but found to be less important. We finally mention the implications with regard to the nature of the deconfinement phase transition, and we discuss the consequences for the formation time of the quark-gluon plasma.

In this chapter we shall investigate the effect of interactions in the quark-gluon plasma. In our previous discussion of the equation of state for the quark-gluon plasma we have treated quarks and gluons as if they were non-interacting particles. The raison d'être of gluons, however, is to mediate the interactions among quarks which leads to permanent confinement of quarks and gluons inside the hadrons (at low temperature). The initial disregard of interactions was based on the property of asymptotic freedom of QCD, i.e. that interactions become weaker and weaker as the distance between particles diminishes. This peculiar feature of QCD known from particle physics carries over to the quark-gluon plasma, as the particle density increases like T^3 and the exchanged momentum grows like T on the average.

At temperatures below the phase transition point, the gluonic interactions among the particles cause the break-up of the quark-gluon plasma phase into the hadronic phase. This process cannot be treated in perturbation theory. At temperatures sufficiently far above the phase transition point, however, one may hope that the interactions are sufficiently weak to render a perturbative expansion meaningful[6]. The lowest order correction to the equation of state of the quark-gluon plasma is proportional to the strong coupling constant α_s. It must be calculated from the same diagrams that led to the 'running' of the coupling constant $\alpha_s(q^2)$, where now expressions for the propagators must be used which describe the corrections due to the presence of the medium. The standard method to treat this problem involves working with propagators periodic in imaginary

[6] A comprehensive review of the present status of perturbative QCD at finite temperature can be found in [Ka83].

time, where the integration over real frequencies is replaced by a sum over the discrete imaginary frequencies [AGD63,FW71]

$$\omega_n = i2n\pi T \qquad \text{for bosons}$$

$$\omega_n = i(2n+1)\pi T \qquad \text{for fermions.} \tag{8.1}$$

For the baryon asymmetric plasma, also a chemical potential μ is introduced for the quarks. The Feynman diagrams in lowest order of perturbation theory are:

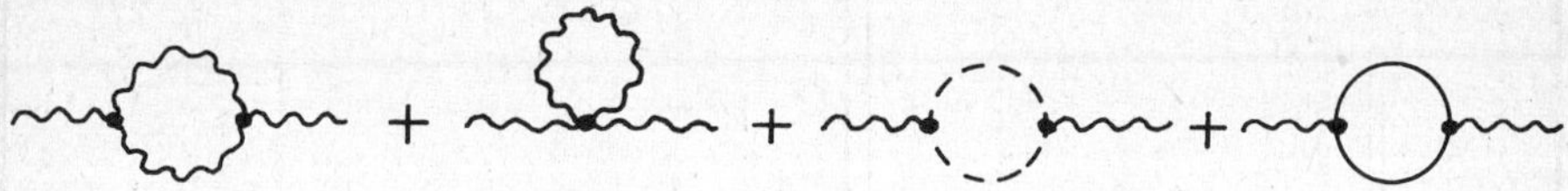

Because the rest system of the quark-gluon plasma introduces a preferred frame of reference, the coupling constant in the presence of the plasma is not only a function of the square of four-momentum transfer, q^2, but also of the time-like component q^0 as a second independent variable. To make things simple, we first ask: what is the coupling constant that describes the interaction among static quarks (or gluons) imbedded in the plasma[7]. We are thus interested in $\alpha_s(|q|=q,q^0=0)$, where q, from now on, denotes only the spatial component of the four-momentum exchange.

For the moment, let us leave aside the contribution of the quark loop diagram. Straightforward evaluation of the Feynman diagrams in the Coulomb gauge yield the following polarization function for loops involving gluons only [Du76]:

---

7 We know by now that quarks and gluons, being massless, move almost with the speed of light, so retardation and magnetic effects must be important for a quantitative description.

$$\Pi(q,T) - \Pi(\mu,0) = (gq/\pi)^2 \, [-11/16 \, \ell n(q^2/\mu^2) + 3\, G(q/T)] \tag{8.2}$$

where

$$G(\xi) = {}_0\int^\infty zdz(e^{\xi z}-1)^{-1} f(z)$$

$$\text{with} \quad f(z) = (z-1/2z+1/8z^3)\ell n(|1-2z|/|1+2z|) - 1 + 1/2z^2 \tag{8.3}$$

constitutes the contribution from the medium. The vacuum contribution, which is temperature-independent, has been renormalized to vanish at some momentum $q_{ren}=\mu$. The effective 'running' coupling constant is obtained by summing all diagrams with a chain of gluon loops:

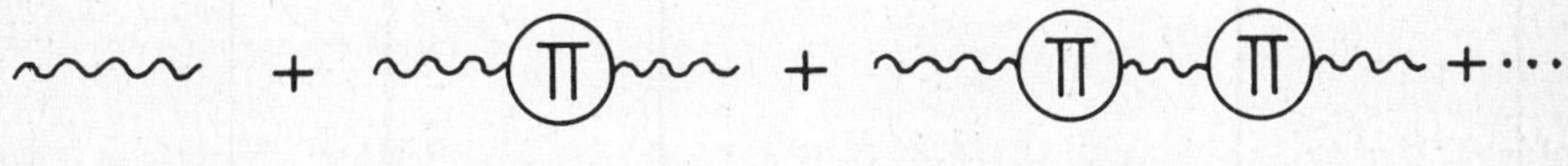

This yields a geometric series that may be resummed into

$$\alpha_s(q,T) = g^2[4\pi(1-\Pi(q,T)/q^2)]^{-1} = 4\pi[11\ell n(q^2/\Lambda^2) - 48G(q/T)]^{-1} , \tag{8.4}$$

where the length scale Λ is a combination of the bare coupling constant and the renormalization point μ.

Before we discuss the momentum dependence of $\alpha_s(q,T)$, it is instructive to look at the zero-momentum limit which governs the long-range behaviour of the gluon propagator in the medium. In the limit $q\to 0$ one finds analytically [Sh78,Sh80,KK80,KK84]:

$$m_{el} = - \mathrm{Re}\, \Pi(0,T) = (gT)^2 - 3/\pi \, (g^2/2)^{3/2} . \tag{8.5}$$

The finite limit of $\mathrm{Re}(\Pi)$ means that the Coulomb propagator $D(q,T) = [q^2-\Pi(q,T)]^{-1}$ contains a mass term[8] in the infrared, reminiscent of the

Debye-Hueckel theory of screening in an electrolytic medium. We now see that the colour charges present in the quark-gluon plasma are polarized in the vicinity of a colour charge in the same way as electric charges are in an electrolyte. In other words: in the quark-gluon plasma colour charges are effectively screened at distances $\Delta x \simeq \pi/q > \pi/gT \equiv \ell_{scr}$. It is just this property which is responsible for the deconfinement phase transition: the quark-gluon plasma is stable at high temperature because the attractive long-range force between coloured particles is screened by the medium[9].

The full momentum dependence of $\alpha_s(q,T)$ requires the numerical evaluation of the integral (3). The result, shown in Fig. 31a , supports the conclusions already drawn from the low-momentum limit. At small momenta ($q<2T$) the effective coupling constant rapidly approaches zero. The coupling constant peaks around $q \simeq 3T$ and becomes virtually indistinguishable from the 'running' coupling constant in vacuo for momenta $q>5T$. The maximal value of $\alpha_s(q,T)$ grows very rapidly as T/Λ approaches the value 0.33, indicating the transition to the confined phase. This behaviour provides an additional estimate of the phase transition temperature for pure gluonic matter, viz. $T_c \simeq \Lambda/\pi \simeq 170$ MeV. Here we have used the value $\Lambda \simeq$ 500 MeV obtained from fits involving space-like momenta, e.g. of quarkonium spectra. Similar results have been obtained by Morely and Kislinger and by Kapusta for the case $T=0$, $\mu \neq 0$ [MK79,Ka79a].

---

[8] The reader should not conclude from this result that the gluon becomes massive in the medium, a result that would invalidate many of our thermodynamic estimates which were based on essentially massless gluons. The mass term (5) is present only in the Coulomb propagator, whereas the propagator of the physical transverse modes remains massless - to this order [Go82,KK82,To82,Ka83].

[9] For small momenta the dynamic Coulomb propagator $D(q^0,q,T)$ behaves like the propagator of massive particles with mass $m=gT$. It describes long-range plasma oscillations called plasmons.

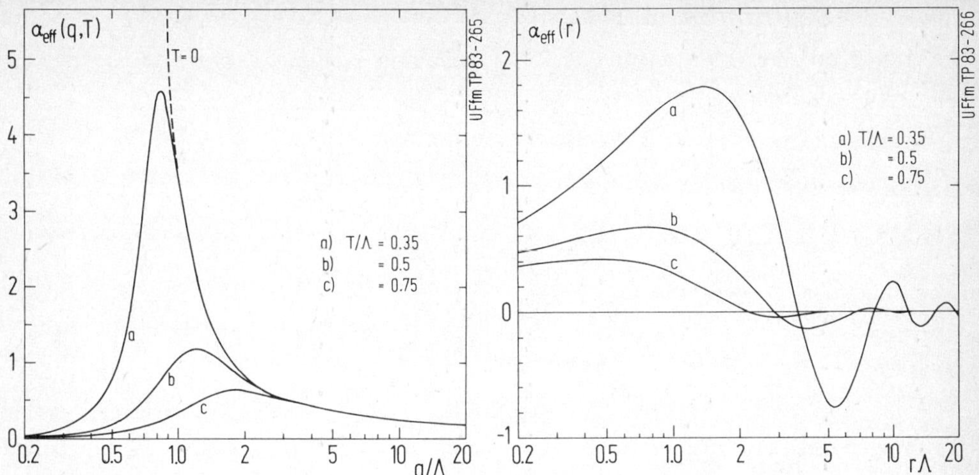

Fig. 31: Effective QCD coupling constant at finite temperature: (a) momentum dependence, (b) distance dependence. The constant Λ has a value of about 500 MeV. Only contributions from gluon loops are taken into account.

It is very instructive to construct the space-dependence of the potential between two static colour charges in the quark-gluon plasma by applying a Fourier transformation to $\alpha_s(q,T)$:

$$V(r) = \alpha_s(r)/r = \int d^3q/(2\pi)^3 \, 4\pi\alpha_s(q,T)/q^2 e^{-iqr} =$$
$$= 2/\pi r \int_0^\infty dq/q \; \alpha_s(q,T)\sin(qr) \; . \tag{8.6}$$

The result is shown in Fig.31b, where one can clearly see that the coupling is screened at large distances beyond $r \simeq 2/\Lambda \simeq 1$fm. This implies that quark-antiquark bound states of larger size cannot exist inside the quark-gluon plasma. E.g. a bound state of a strange quark-antiquark pair (a "ϕ-meson") is expected to have a diameter $r_{ss} \simeq (\alpha m_s/2)^{-1} \simeq 5$fm for $\alpha_s \simeq 0.5$ and $m_s \simeq 150$MeV. On the contrary, a charmonium state could be stable, because $r_{cc} \simeq (\alpha m_c/2)^{-1} \simeq 0.5$fm for $m_c \simeq 1500$MeV.

The contribution of the interactions to the equation of state of the quark-gluon plasma involves a weighted average over $\alpha(q,T)$, sampling no t only space-like momenta but also time-like momenta because the coloured constituents of the plasma a fast-moving, essentially massless particles.

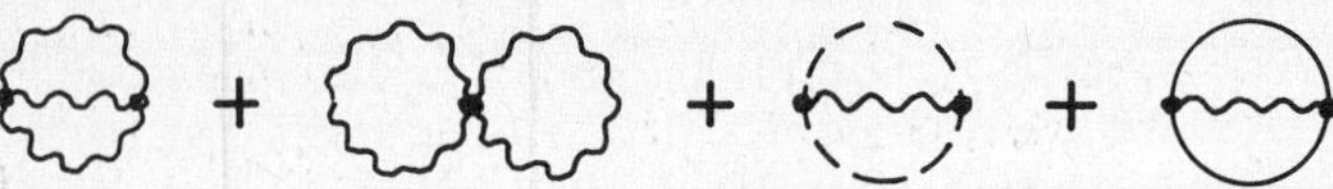

Fig.32: Feynman diagrams contributing to the equation of state of the quark-gluon plasma in order α_s.

In lowest order, four diagrams contribute to the equation of state (see Fig.32): the gluonic self-energy diagrams (a,c), the ghost loop diagram (b), that cancels the unphysical glue modes, and the quark self-energy (d). Evaluating these Feynman diagrams, one finds that the energy density, the pressure and baryon density of the quark-gluon plasma, expressed in terms of the intensive variables T and μ, are decreased according to [Ch78,KK79,Ka79,Ka83a]

$$\varepsilon = 3\,P = (1-15\alpha_s/4\pi)8\pi^2T^4/15 + (1-50\alpha_s/21\pi)7\pi^2T^4/10 + (1-2\alpha_s/\pi)(\pi^2T^2+\mu^2/2)3\mu^2/\pi^2 \;,$$

$$n_b = 2/3\pi^2\,(1-2\alpha_s/\pi)\mu(\pi^2T^2+\mu^2)\;. \tag{8.7}$$

The first term in ε contains the gluonic contribution, the remaining terms account for the contributions from quarks. Comparison with eq. (3.6) reveals that the number of degrees of freedom in the quark-gluon plasma is effectively reduced by the interaction, at least in lowest order, i.e. for small α_s.

That this does not necessarily remain true for larger values of α_s is shown by the next contribution to the gluonic energy density. Contrary to naive expectations[10], it turns out to be of the order $\alpha_s^{3/2}$, and is obtained by summing all ring diagrams including the polarization operator (2) at least twice:

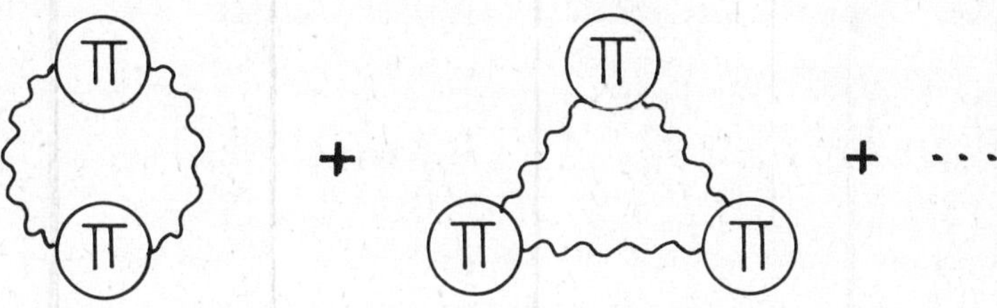

Fig.33: Summation of an infinite series of loop diagrams to obtain the plasmon contribution to the energy density of the quark-gluon plasma.

This so-called plasmon contribution to the energy density of the quark-gluon plasma takes the form

$$E_{Pl} = 2T^2/\pi^2 \; {}_0\!\int^{\infty} dk \; \partial\Pi/\partial T \; \Pi(k)/(k^2-\Pi) \; = \; 2/\pi \; g^3T^4 \; , \tag{8.8}$$

where the last result is obtained by substituting the infrared limit $\Pi=-(gT)^2$ for the polarization function. If the expression (8) is included in the gluonic contribution to the energy density in eq. (7), the coefficient is modified to [Ka79a,To83]

---

[10] The noninteger power of α_s is a sign of the non-perturbative character of this contribution. Since each diagram in Fig. 33 is individually infrared divergent, an infinite number of such diagrams must be added, i.e. contributions from all orders of perturbation theory must be summed to obtain an infrared finite result. As we have seen, the summation of the loop diagrams leads to a finite screening length or, in other words, to a non-zero mass $m=gT=4\pi\sqrt{\alpha_s}T$ for Coulomb gluons. The unusual power of α_s is caused by this mass term.

$$1 - 15\alpha_s/4\pi + 30(\alpha_s/\pi)^{3/2} \; . \tag{8.9}$$

Even for moderately large values of α_s, e.g. $\alpha_s=1/2$, the new contribution is astonishingly large. It may be that a self-consistent calculation of the electric gluon mass, including the plasmon effects, would cure this situation [KT83]. The state of affairs is aggravated by the observation that other diagrams of order α_s^2 and beyond, which are not included in the plasmon contribution, contain gauge-dependent infrared divergences of a still unresolved nature [Na82]. In addition, there are contributions to the energy density of the form $e^{-1/\alpha}$ due to vacuum tunneling processes, the so-called instantons. We have thus to conclude that the inclusion of the effect of quark-gluon interactions in the framework of perturbation theory rests on weak ground.

All properties of a thermodynamic ensemble in thermal equilibrium can be derived from the partition function

$$Z = \mathrm{Tr}(e^{-\beta\hat{H}}) = \Sigma_i \, e^{-\beta E_i} , \tag{9.1}$$

where the sum runs over all states accessible to the system. E.g., the internal energy E and the pressure P are obtained as

$$E = \varepsilon V = T^2 \partial(\ln Z)/\partial T \ , \ P = T\partial(\ln Z)/\partial V. \tag{9.2}$$

For a non-interacting gas of massless bosons or fermions the sum can be done analytically yielding the simple result

$$\ln Z_B = g_B \pi^2/30 \ VT^3 \ , \quad \ln Z_F = g_F 7\pi^2/120 \ VT^3 \ , \tag{9.3}$$

where g counts the respective independent degrees of freedom (for gluons g_B=2×8=16, for quarks g_F=2×3×N_f=6N_f).

The equation of state derived from eq.(3) formed the basis for our previous estimates of the quark-gluon plasma. We have also investigated the effect of the interaction among the coloured particles in perturbation theory. But we have not yet considered the most dramatic effect of the QCD interaction: that only states can exist which are global colour singlets! Although colour confinement is broken locally in the quark-gluon plasma, and "long-range" colour fluctuations are possible throughout the whole region where nuclear matter is in the plasma phase, the plasma fireball taken as a whole must remain a colour singlet. The transition from hadronic matter to the quark-gluon plasma phase is a transition from local colour confinement (on the scale of 1 fm) to global colour confinement.

This means that only states forming a global colour singlet are accessible, and therefore the sum in eq.(1) must be restricted to states without open colour. In calculating the equation of state for the quark-gluon plasma we are really interested in the restricted partition fuction

$$Z_1 = \mathrm{Tr}_{(1)}[e^{-\beta\hat{H}}] , \tag{9.4}$$

where the index "1" means restriction to states transforming under the colour singlet representation.

To attack the problem by explicitly constructing many-particle states of singlet symmetry according to the rules of SU(3) group theory probably would be a futile attempt. Too many different ways exist to couple n particles to a given SU(3) symmetry. Fortunately, there exists an indirect method, where one constructs a generating partition function from which the restricted partition function for any given irreducible representation can be obtained by projecting with the appropriate invariant group function (the character function).

To see how the method works, we first study a simple example. Consider a gas of fermions of a single species (particles and antiparticles) which possess a conserved, additive quantum number n (e.g. baryon number). Now let us find the partition function restricted to a specific total value of $n = n_0$:

$$Z_n = \mathrm{Tr}_{(n)}[e^{-\beta\hat{H}}] . \tag{9.5}$$

Observe that n_0 can be made up from (n_0+m) particles and m antiparticles with any value of m!

The usual way of treating this problem in a crude manner is to define the grand partition function

$$Z(\mu) = \mathrm{Tr}[e^{-\beta(\hat{H}-\mu\hat{N})}] = \sum_n \mathrm{Tr}_{(n)}[e^{-\beta(\hat{H}-\mu n)}] = \Sigma\, e^{n\beta\mu} Z_n . \tag{9.6}$$

One then fixes the chemical potential μ by equating the average value of the quantum number with the required value n_0:

$$n_0 = \langle\hat{N}\rangle = Z(\mu)^{-1} \Sigma\, n e^{n\beta\mu} Z_n = T\, \partial[\ln Z(\mu)]/\partial\mu . \tag{9.7}$$

This method, however, works satisfactorily only when the total number of particles is very large and fluctuations in the quantum number can be safely neglected.

Surprisingly, the exact solution can be obtained with little additional effort. Introducing the fugacity $\lambda = e^{\beta\mu}$, the expression (6) becomes a Laurent series in the variable λ:

$$Z(\lambda) = \Sigma_n \lambda^n Z_n . \tag{9.8}$$

We now may obtain any coefficient Z_n in this series by Cauchy's integral formula, if we integrate around the origin in the λ-plane:

$$Z_n = \oint d\lambda/2\pi i\; Z(\lambda) e^{-(n+1)} . \tag{9.9}$$

Parametrizing the integration path by $\lambda = e^{i\phi}$ $(0<\phi<2\pi)$, we are left with the projection formula

$$Z_n = \int d\phi/2\pi\; e^{-in\phi} Z(\phi) , \tag{9.10}$$

where

$$Z(\phi) = Z(\lambda = e^{i\phi}) = \mathrm{Tr}[e^{-\beta\hat{H}+i\phi\hat{N}}] \tag{9.11}$$

is the generating function for all restricted partition functions Z_n.

The symmetry group for an additive quantum number, such as charge, baryon number, etc., is the abelian Lie group U(1). All that is left to do is to generalize the projection method to non-abelian Lie groups, such as the colour group SU(3). This was done by Redlich and Turko [RT80,Tu81], whose method we outline in the following. Assume that the symmetry to be imposed on the partition fuction is described by a (compact) Lie group. Denote the irreducible representations by an index C, and label the members of each of these multiplets by an additional index $\alpha = 1,\ldots,d_C$, where d_C is the dimension of the multiplet C. (In practice one will use the eigenvalues of the Casimir operators to specify the multiplet and label the members of a multiplet by the eigenvalues of the generators in the Cartan subalgebra, i.e. the maximal commuting subalgebra.) Thus we have

$$Z = \sum_C \mathrm{Tr}_{(C)}[e^{-\beta\hat{H}}] \equiv \sum_C Z_C = \sum_C (\sum_\alpha Z_{C\alpha}) . \tag{9.12}$$

If the symmetry is unbroken, all members of a given multiplet have the same energy, hence

$$Z_{C\alpha} = Z_C/d_C \qquad \text{for all } \alpha. \tag{9.13}$$

To find the restricted partition function

$$Z_{C^0} = \mathrm{Tr}_{(C^0)}[e^{-\beta\hat{H}}] = \sum_{C^0} Z_C \delta_{C,C^0}. \tag{9.14}$$

we need a suitable representation of the Kronecker symbol. This representation is provided by the character functions of the Lie group (restricted to the maximal abelian subgroup):

$$\chi_C(\phi_\nu) = \sum_\alpha \langle C,\alpha|\ \exp(i\sum_\nu \phi_\nu \hat{I}_\nu)\ |C,\alpha\rangle . \tag{9.15}$$

Here $|C,\alpha\rangle$ denotes the states in the multiplet labelled C, and the sum over ν runs over all commuting generators $\hat{I}_\nu$, their number being equal to the rank of the Lie group. The character functions satisfy the orthogonality relation

$$\delta_{C,C'} = d_C^{-1} \int d\mu(\phi_\nu)\, \chi^*_{C'}(\phi_\nu)\chi_C(\phi_\nu) \tag{9.16}$$

when the integration is carried out with respect to the invariant (Haar) measure $\mu(\phi)$ of the continuous group.

The important property of the definition (15) for the invariant group function is that it is independent of the microscopic structure of the states $|C,\alpha\rangle$ which transform under the irreducible representation C, i.e. it does not matter of how many particles the multiplet is made up. Combining eqs.(13-15) we therefore have

$$\begin{aligned} Z(\phi_\nu) &= \Sigma_C\, d_C^{-1} Z_C \chi_C(\phi) = \sum_{C\alpha} Z_{C\alpha} \langle C\alpha|\exp(i\Sigma\phi_\nu \hat{I}_\nu)|C\alpha\rangle \\ &= \sum_{C\alpha} \langle C\alpha|e^{-\beta\hat{H}} e^{-\Sigma\phi_\nu \hat{I}_\nu}|c\alpha\rangle = \mathrm{Tr}\,[\exp(-\beta\hat{H}+i\Sigma\phi_\nu\hat{I}_\nu)] \quad . \end{aligned} \tag{9.17}$$

Eqs.(16) and (17) are the basic equations which permit the calculation of partition fuctions for systems restricted to any symmetry. The computational simplicity derives from the fact that the operators in the exponential in eq.(17) have additive eigenvalues for a system of non-interacting particles, allowing for analytic evaluation of $Z(\phi)$ in the absence of interactions. For interacting particles one can derive a perturbative (diagrammatic) expansion for $Z(\phi)$ by standard methods. Here we shall neglect interactions and simply explore the fact of the symmetry restriction[11].

---

[11] Neglecting perturbative interactions may not be so bad for the quark-gluon plasma. The most drastic conseqence of the colour interaction is to cause global confinement of coloured particles, and we paid tribute to precisely this effect by restricting the partition function to colour singlet states. A large part of the

For the abelian group U(1) considered before all irreducible representations are one-dimensional. They are labelled by an integer n, their character functions being $\chi_n(\phi)=e^{in\phi}$. With the measure $d\mu(\phi)=d\phi/2\pi$, the eqs.(16,17) reduce to the result (10,11) which we had derived the pedestrian way, without the help of group theory. For the simplest non-abelian group SU(2) the irreducible representations are labelled by a half-integer $j=0,\frac{1}{2},1,\ldots$ with dimension $d_j=2j+1$. The character functions are found to be [Gi74]

$$\chi_j(\phi) = [\sin\tfrac{1}{2}(2j+1)\phi]/\sin(\tfrac{1}{2}\phi) \ , \tag{9.18}$$

while the invariant group measure is

$$d\mu(\phi) = (\sin^2\tfrac{1}{2}\phi)d\phi/2\pi \text{ with } 0<\phi<4\pi \tag{9.19}$$

For the colour group SU(3), which has rank 2, the irreducible representations are usually labelled by two integers p and q, and we need two angles ϕ_1 and ϕ_2 to parametrize the elements of the maximal abelian subgroup. The general expressions for the characters of SU(3) and the group measure are fairly complicated and we will not write them down explicitly here. Instead, we refer the reader to the work of Elze et al. [EGR83] and Gorenstein et al. [Go83].

In order to investigate the effect of symmetry restriction on the equation of state of the quark-gluon plasma and to see how the method works in practice, we shall treat a simplified model in some detail: we replace the colour group SU(3) by the simpler non-abelian group SU(2), and we assume that the overall baryon number is zero. This is usually described by setting the chemical potential for the quarks $\mu=0$, but now we can fix the

---

non-perturbative aspects of the QCD interaction is thus contained in our result.

baryon number exactly by projection. Let us further assume that there are N_F quark flavours. In our SU(2) model, gluons come in three colours, quarks and antiquarks each in two. We need one angle θ for the colour projection and another angle ϕ for the projection on baryon number zero for each quark flavour. The partition function restricted to colour singlet states with zero baryon number is then given by

$$Z_{1,0} = \mathrm{Tr}_{(1,0)}(e^{-\beta\hat{H}}) = \int d\theta/2\pi \sin^2\tfrac{1}{2}\theta \, [\int d\phi/2\pi \, Z(\theta,\phi)]^{N_F} , \tag{9.20}$$

since the character functions are identically 1 for the simplest representations. The generating function can be factorized into the contributions from gluons, quarks and antiquarks:

$$\begin{aligned} Z(\theta,\phi) &= \mathrm{Tr}\,[e^{-\beta\hat{H}+i\theta\hat{I}_3+i\phi\hat{B}}] = Z_G(\theta)Z_Q(\theta,\phi)Z_{\bar{Q}}(\theta,\phi) = \\ &= Z_G(\theta)|Z_Q(\theta,\phi)|^2 \end{aligned} \tag{9.21}$$

where $\hat{I}_3$ is the third generator of colour-SU(2) and the operator B takes the eigenvalues +1 for quarks and -1 for antiquarks. The last identity in eq.(21) holds because $Z_{\bar{Q}} = Z_Q^*$. The eigenvalues of $\hat{I}_3$ are -1,0,+1 for the gluons and $\pm\frac{1}{2}$ for the quarks and antiquarks.

For non-interacting quarks and gluons the energy of a many-particle state is given by the sum of the single-particle energies ω. We therefore find the following expressions:

$$\begin{aligned} Z_G(\theta)^{-1} &= \prod_\omega (1-e^{-\beta\omega+i\theta})(1-e^{-\beta\omega})(1-e^{-\beta\omega-i\theta}) \\ Z_Q(\theta,\phi) &= \prod_\omega (1+e^{-\beta\omega+i\frac{1}{2}\theta+i\phi})(1+e^{-\beta\omega-i\frac{1}{2}\theta+i\phi}) \end{aligned} \tag{9.22}$$

A convenient way of evaluating the infinite products is by taking the logarithm on both sides and expanding the individual logarithms for each mode into power series. After some algebra one finds

$$\ln Z_G(\theta) \;=\; \Sigma_n\, n^{-1}\zeta_n^G(3-4\sin^2\tfrac{1}{2}n\theta)$$

$$\ln|Z_Q(\theta,\phi)|^2 \;=\; \Sigma_n\, (-)^{n-1}n^{-1}\zeta_n^Q[4-8\sin^2 n\theta/4-8\sin^2(\tfrac{1}{2}n\phi)\cos\tfrac{1}{2}n\theta] \;. \tag{9.23}$$

Here $\zeta_n^{G/Q}$ denotes a kind of single-particle partition function for gluons/quarks:

$$\zeta_n \;=\; \Sigma_\omega\, e^{-n\beta\omega} \;. \tag{9.24}$$

In the continuum limit, the mode sum in eq.(24) becomes an integral over momentum space, which yields for massless particles

$$\zeta_n^G \;=\; \zeta_n^Q \;=\; 2VT^3/\pi^2 n^3 \quad \text{for } VT^3\to\infty. \tag{9.25}$$

Exponentiating the expressions (23) and expanding up to quadratic terms in the angles, the angular integrals in eq.(20) can be performed by the method of steepest descent. The result is (in the limit $VT^3\to\infty$):

$$\ln Z_{1,0} \;\simeq\; (6+7N_F)\pi^2/90\; VT^3 \;-\; \tfrac{1}{2}(3+N_F)\;\ell n(VT^3) \;+\; \text{const.} \tag{9.26}$$

The first term is the one obtained without restriction to colour symmetry, the second term yields the wanted correction. From this result we can now compute the energy density and pressure of the colour singlet quark-gluon plasma with the help of the formula (2). One conclusion can be drawn without further algebra: since $Z_{1,0}$ contains the volume V and the temperature T only in the combination VT^3, the relation E=3PV, which is generally true for massless particles, continues to hold.

A useful way of plotting the result is to take the ratio D_{eff} of the energy density $\varepsilon_{1,0}$, as calculated from eq.(26), to the energy density ε obtained without restriction to colour symmetry. The latter is given by the first term in eq.(26). This ratio describes the effective reduction in the number of degrees of freedom caused by the requirement of global colour neutrality. In the limit of VT^3 large we find:

$$D_{eff} = E_{1,0}/E = 1 - 45/\pi^2 \, (3+N_F)/(6+7N_F) \, (VT^3)^{-1}. \tag{9.27}$$

To obtain a feeling where the correction becomes important, we may ask where the number of degrees of freedom is reduced by half, i.e. where $D_{eff}=\frac{1}{2}$. For $N_F=0$ we find $VT^3 \simeq 5.6$, and $VT^3 \simeq 2.3$ for $N_F=2$. When the angular integrals in eq.(20) are performed numerically, one finds for the continuum limit (25) the result shown in Fig. 34 by the solid line.

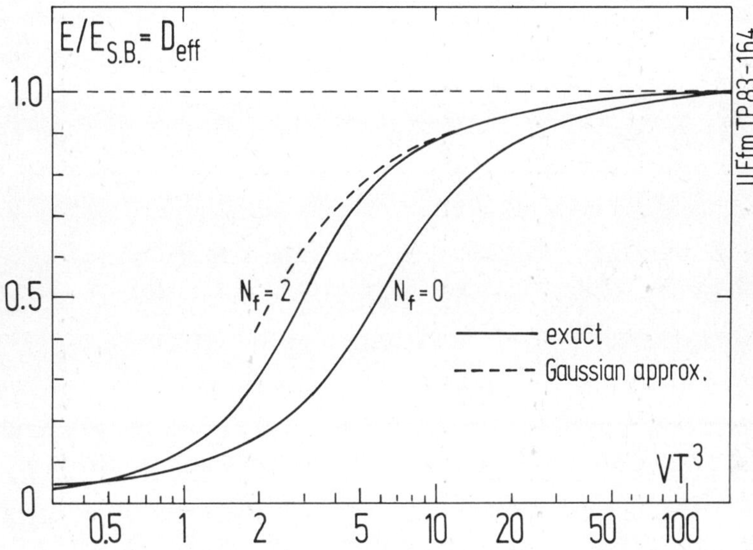

Fig. 34: Effective reduction of the number of degrees of freedom in SU(2) gauge theory because of the requirement of global colour symmetry, for zero and two quark flavours.

As can be seen, the reduction of $D_{eff.}$ is slightly stronger than estimated by eq.(27). The steep transition from $D_{eff} \simeq 1$ to $D_{eff} \simeq 0$ is almost reminiscent of a phase transition, but an inspection of the specific heat shows that there is no discontinuity involved.

Elze, Greiner and Rafelski [EGR83] have performed the same calculation for the true colour group SU(3) with two quark flavours. Their result, shown in Fig.(35) shows the same steep transition region where most of the degrees of freedom of the quark-gluon plasma are frozen in within a small range of values of the variable VT^3. Since there is still a ten percent reduction at $VT^3=10$, the colour restriction must be taken into account for small and medium sized regions of quark-gluon plasma. The influence of a finite chemical potential μ of the quarks has been studied in [EGR84].

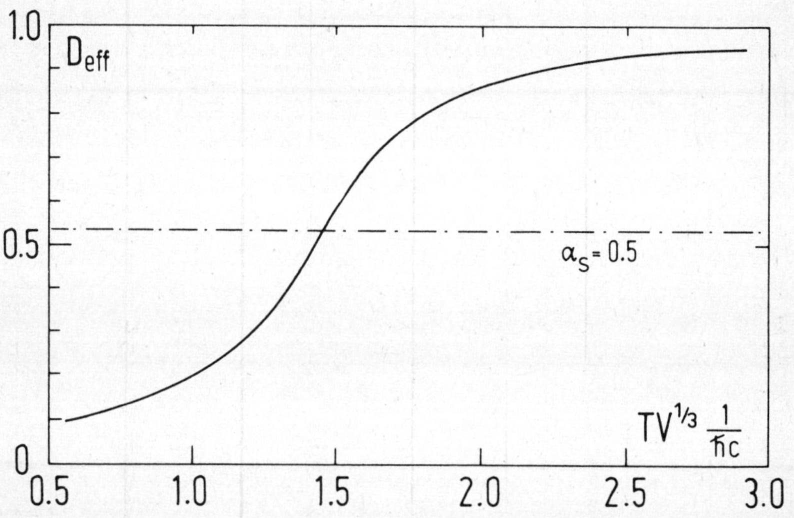

Fig. 35: Effective reduction in the number of degrees of freedom in SU(3) gauge theory (true QCD) with two quark flavours (from [EGR83]).

Since the symmetry correction (27) has the dependence V^{-1} for large volumes, it can be understood as a correction due to finite particle number in the volume. As such it forms only one part of the general expansion of the partition function for finite volume which is known to be a power series in the parameter $V^{-1/3}$:

$$Z(V) = Z_0 + Z_1 V^{-1/3} + Z_2 V^{-2/3} + Z_3 V^{-1} + \ldots \tag{9.28}$$

Only the term Z_0 survives in the so-called "thermodynamic" limit $V\to\infty$. The next term, Z_1, is the surface correction, Z_2 describes the effects due to the finite diameter of the volume, and it is only Z_3 which contains - among other things - corrections due to finite particle number. How do we know, then, that the symmetry correction is more important than other finite volume corrections?

The complete function Z(V) can be obtained if one uses the exact discrete single-particle energies in the mode sum (24) instead of the continuum approximation. When we discussed the MIT-bag model in ch.2, we wrote down the conditions for the eigenmodes of massless quarks and gluons confined to a spherical cavity of radius R. The eigenvalue equations (2,4) and (2,11) can be resolved with the help of the asymptotic expansion of the spherical Bessel function

$$z\, j_\ell(z) \simeq \sin(z-\ell\pi/2) + \ell(\ell+1)\cos(z-\ell\pi/2)/2z \, . \tag{9.29}$$

In this way the following approximate energy eigenvalues are obtained (k=0,1,2,...):

$$\begin{aligned} &\text{quarks:} \quad \omega^Q_{k\ell} R = x^Q_{k\ell} = (\ell+1+k)\pi/2 - (\ell+\tfrac{1}{2})^2/(\ell+1+k)\pi \quad (\ell=\tfrac{1}{2}, 1\tfrac{1}{2}, \ldots) \\ &\text{gluons:} \quad \omega^G_{k\ell} R = x^G_{k\ell} = (\ell+1+k)\pi/2 - \ell(\ell+1)/(\ell+1+k)\pi \quad (\ell=1,2,3,\ldots) \end{aligned} \tag{9.30}$$

It is found that the deviation from the exact eigenvalues is less than ten percent in alll cases. The mode sums (24) are now functions of the dimensionless variable $RT = (3V/4\pi)^{1/3}T$:

$$\zeta_n(RT) = \Sigma_\ell \, (2\ell+1) \, \Sigma_k \, e^{-nx_{k\ell}/RT} \, . \tag{9.31}$$

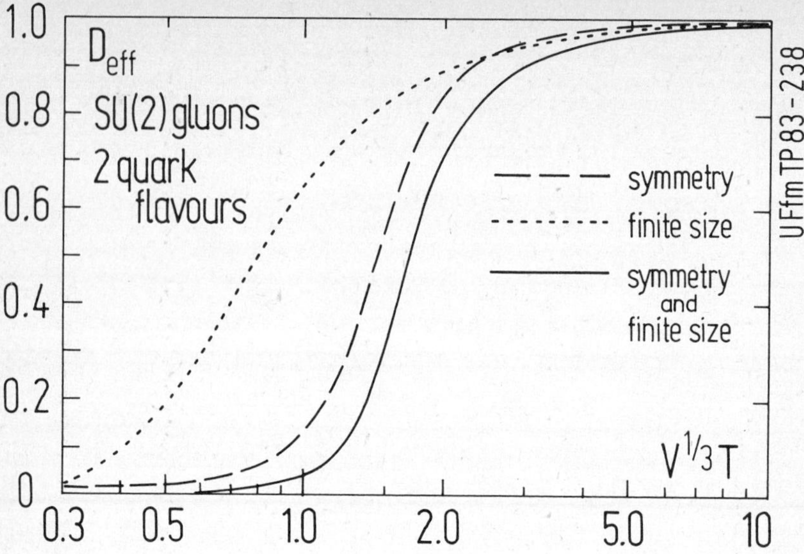

Fig. 36: Effective reduction of the number of degrees of freedom for SU(2) gauge theory with global colour symmetry and finite size corrections. The Casimir energy is not included.

Numerical evaluation of the colour-restricted partition function $Z_{1,0}(TV^{1/3})$ with the discrete eigenmodes yield the result shown by the solid line in Fig.36 (for two flavours and the simplified SU(2) colour group). For comparison we have also included in the figure the suppression factor without colour restriction but with the discrete modes[12] (dotted line) and the colour-restricted suppression factor in the continuum limit (dashed curve), which was already shown in Fig.34. Although the pure finite size effects are seen to dominate at large values of $TV^{1/3}$, the

---

[12] In principle, the finite size calculation should also include the Casimir energy, i.e. the modification of the vacuum energy due to mode discretization (see e.g. [PHJ82]). There is a partial cancellation of the thermal effects in the high-temperature limit [Me68,BM69], as can be shown quite generally. Unfortunately, the Casimir energy of the quark field has pathological features in the MIT-bag model (see

main drop in the apparent number of degrees of freedom is almost entirely due to colour symmetry restriction. The main influence of the finite size is to shift the transition region to larger values of $TV^{1/3}$. For small values of $TV^{1/3}$ the solid line approaches zero rapidly, because the discretization of the eigenmodes introduces a gap in the excitation spectrum of quarks and gluons.

But let us go back to Fig. 35, which showed the effective suppression of degrees of freedom for SU(3) due to global colour symmetry. In addition, this figure contains the reduction factor caused by perturbative colour interactions for $\alpha_s=\frac{1}{2}$ according to eq. (8.7). Now we know from ch. 1 that the effective "running" coupling constant $\alpha_s(r)$ increases with distance, i.e. with the size of the volume. The reduction due to α_s is therefore increasing with the volume V. Obviously the two effects, i.e. global colour symmetry and perturbative interaction, work against each other.

Under given external conditions, here for given temperature T, a thermodynamic system tries to maximize its entropy, which means that it tries to assume a state with the largest number of accessible degrees of freedom. Therefore, in the light of our discussion above, the global colour symmetry tries to increase the confinement volume, whereas the perturbative interactions try to reduce the size of the volume. In general, the balance will be reached at some finite volume $V_{eq}(T)$. At zero temperature V_{eq} corresponds to the normal hadronic volume as described, e.g., in the bag model.

Above the critical temperature T_c, the long-range part of α_s is screened due to the presence of the polarizable medium, as we have seen in ch. 8 (see Fig. 31). The perturbative interactions then no longer counteract the global colour symmetry effect for large volumes, and the colour correlation volume will try to become as large as the extensions of the system

---

[Mi83,BI83]). We have, therefore, not considered the Casimir energy here.

permit. In order to obtain quantitative understanding, the partition function $Z_{1,0}$ must be calculated in the presence of perturbative interactions. This may lead to an interesting model for the colour deconfinement phase transition.

Let us finally discuss which implications the effective dependence of the number of degrees of freedom has for the formation of the quark-gluon plasma. As we just said, the transition from the hadronic phase to the quark-gluon plasma can be understood - crudely speaking - as a change in the volume over which colour is confined. The hadronic phase corresponds to colour confinement on a scale of about 1 fm, whereas the size of the confinement volume is much large in the plasma phase. In other words, a quark-gluon plasma with a confinement range of 1 fm is, effectively, a gas of hadrons.

Looking at Fig. 35 we find that the variable $V^{1/3}T/hc$ must grow in value from about 1 to 2 for D_{eff} to jump from 0.1 to 0.9. As we have learned, finite size effects can be expected to shift these points by about 0.5 in the same direction. To increase the confinement volume, therefore, the range of colour equilibration must be increased by $\Delta(V^{1/3}) \simeq 1$ fm, where we used T = 200 MeV. This corresponds to a change of $\Delta R \simeq 0.6$ fm in the radius of the confinement volume.

It is clear that this cannot occur instantaneously, since the particles that carry colour cannot propagate faster than light. Our argument leads us, therefore, to an estimate for the time required to make the transition from hadronic to quark matter. We may call this the formation time τ_f of the quark-gluon plasma, our estimate being $\tau_f \simeq \frac{1}{2}$ fm/c = 1.5×10^{-24} s. Whether this lower limit is attained in reality depends on the cross-sections for the mechanisms leading to colour transport. The rate for one such process, the conversion of gluons into light quarks or vice versa ($gg\rightarrow q\bar{q}$) is calculated in ch. 12, where we find a value of $1\text{-}2\times10^{-24}$ s in accord with our other estimate. We conclude that the formation time of the quark-gluon plasma is probably about $\frac{1}{2}$ fm/c.

LECTURE 5

The fifth and final lecture is devoted to the internal dynamical evolution of the quark-gluon plasma. We begin with a discussion of the transport theory of the plasma, starting from an ensemble of classical, pointlike coloured particles interacting through an average colour field. We introduce a phase-space distribution function f(x,p,Q) which satisfies a transport equation of the Vlasov type. We then rederive these equations within the framework of quantum mechanics, where the Wigner function takes the role of the classical distribution function, and obtain a set of hydrodynamical equations for the quark-gluon plasma.

We then investigate the relativistic hydrodynamical equations neglecting collective colour currents. Assuming Lorentz invariant initial conditions, as they may be realized in the central rapidity region, a simple scaling solution of the hydrodynamical equations can be found. In this solution, the plasma volume expands mainly by longitudinal growth, resulting in a rapid decrease of the energy density.

In the final chapter we address the question raised in ch.5, whether there is sufficient time to reach chemical equilibrium for strange quarks during the existence of the plasma state. Calculating the rate of strange quark-antiquark pair production in lowest order of perturbation theory, we obtain the surprising result that strangeness production is completely dominated by pure gluonic production mechanisms. This tells us that strangeness - as opposed to electromagnetic signals - is a sensitive trigger for the gluon content of the quark-gluon plasma. Solving the rate equation for strangeness production, we find a saturation time of $2\text{-}3\times10^{-23}$ s or 6-10 fm/c.

To describe the evolution of the quark-gluon plasma created in a high-energy nuclear collision, a theory of transport phenomena in the plasma of coloured particles is needed. An attempt to formulate such a theory has been started by U.Heinz [He82,83] and also by McLerran [LM83]. Heinz has suggested that the structure of the equations can be obtained by considering an ensemble of classical, structureless coloured particles in interaction with an average local colour field $A_\mu^a(x)$.

If the world lines of the particles are parametrized by the proper time τ, $\xi_i^\mu(\tau)$, and their momenta are given by $p_i^\mu(\tau)$, the equations of motion are [Wo70,Ar82]:

$$dp_i^\mu/d\tau \quad = \quad Q_i^a(\tau)\ F_a^{\mu\nu}[\xi(\tau)]\ d\xi_\nu^i/d\tau \quad . \tag{10.1}$$

Here it is assumed that the charge vector $Q^a(\tau)$ is in the adjoint representation of SU(3), having eight components. Because the mean field itself can carry colour, the particles may exchange colour with the field.

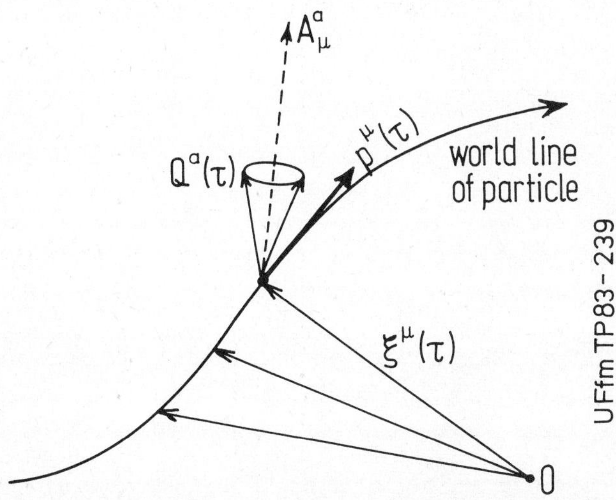

One can show that this leads to a precession of the colour charge according to

$$dQ_i^a \quad = \quad f_{abc} \, d\xi_i^\mu/d\tau \; A_\mu^b[\xi(\tau)] \; Q_i^c(\tau) \ , \tag{10.2}$$

which conserves the two Casimir invariants of SU(3), Q^aQ_a and $d_{abc}Q^aQ^bQ^c$. Self-consistency of the equations is obtained, if one requires that the mean field is created by the coloured particles themselves, i. e.

$$\partial_\mu F_a^{\mu\nu} + f_{abc} A_\mu^b F_c^{\mu\nu} \quad = \quad j_a^\nu \tag{10.3}$$

where

$$j_a^\nu(x) \quad = \quad \sum_i \int d\tau \; \delta[x-\xi(\tau)] \; Q_i^a(\tau) \; d\xi_i^\nu/d\tau \ . \tag{10.4}$$

In order to make the transition to transport equations for a continuous medium, one introduces a one-particle distribution function $f(x^\mu, p^\nu, Q^a)$ describing the probability to find a particle at space-time point x^μ with momentum p^ν and colour charge vector Q^a. Liouville's theorem states that the density of particles in phase space stays the same along the world lines of the particles, which means that $df/d\tau = 0$. This equation holds as long as no two-body collisions occur in the quark-gluon plasma. Carrying out the differentiation and making use of the equations of motion (1,2) the transport equation

$$p^\mu \partial f/\partial x^\mu + Q_a F_a^{\mu\nu} p_\nu \partial f/\partial p^\mu + f_{abc} p^\mu A_\mu^b Q_c \, \partial f/\partial Q^a \quad = \quad 0 \tag{10.5}$$

is obtained. When two-body collisions are included, the right-hand side will contain a collision term describing the loss of particles in the initial states and the particle gain in the final states in phase space. In this way eq.(5) would become a Boltzmann equation for the coloured particle distribution fuction f(x,p,Q).

So far, we have said nothing about the detailed nature of the coloured particles, except that they are point-like. If they are to represent quarks, their colour states should form a SU(3) triplet, they should be Fermions and they should have their respective anti-particles. All these conditions are difficult to formulate in a classical theory, so we shall follow recent work of Heinz [He83] and start directly from the quantum theory, where quarks are described by a colour triplet Dirac field.

The one-body density matrix is given by the expectation value of

$$\hat{F}(x_1,x_2) \quad = \quad :\hat{\bar{\Psi}}(x_1)\, U(x_1,x_2)\, \hat{\Psi}(x_2): \tag{10.6}$$

where the operator

$$U(x_{1,2}) \quad = \quad P \exp[i\tfrac{1}{2}\lambda_a \int dz^\mu A^a_\mu(z)] \tag{10.7}$$

is introduced to make $\hat{F}$ gauge-invariant. The physical meaning of U is to transport the coordinate system in colour space from point x_2 to x_1 in order to allow a gauge-independent comparison between quark fields at different space-time points (the symbol "P" here means path ordering of the operators in the exponential). The analogue of the classical phase space distribution function is obtained by forming the Fourier transform with respect to the relative coordinate $v = x_1 - x_2$. Writing $x = \frac{1}{2}(x_1+x_2)$, the so-called Wigner function (see e.g. [GLW80,CZ83]) is defined by the expectation value of $\hat{F}$ in the ensemble:

$$F(x,p) \quad = \quad \int (dv)\, e^{-ipv} \langle \hat{F}(x+\tfrac{1}{2}v, x-\tfrac{1}{2}v)\rangle \,. \tag{10.8}$$

Although F(x,p) is the analogue of the classical distribution f(x,p,Q), there are some differences. F(x,p) contains contributions from quarks and antiquarks alike, which can be isolated by splitting F - for time-like momenta - into a positive-frequency and a negative-frequency part, F =

$F^{(+)}\theta(p^0) + F^{(-)}\theta(-p^0)$. In addition, F is a 4×4 spin matrix, which is necessary to describe the spin degrees of freedom of the quark fluid. In the same way, the colour degrees of freedom are described by F being a 3×3 matrix in colour space. Physical observables are expressed by the spin and colour trace taken with the adequate operators. For instance, the current density of the quark fluid is

$$j_a^\mu(x) = \int (dp)\, \mathrm{Tr}[\tfrac{1}{2}\lambda_a \gamma^\mu F(x,p)] \,, \tag{10.9}$$

and the energy momentum tensor is given by the expression

$$T^{\mu\nu} = \int (dp)\, p^\mu\, \mathrm{Tr}[\gamma^\nu F(x,p)] \,. \tag{10.10}$$

The difference, i.e. that the classical function f(x,p,Q) is a function of the continuous vector Q^a, whereas the quantum distribution function F(x,p) has a finite number of colour components, reflects the quantal nature of the colour variable: quarks are in the colour triplet representation, and their evolution in colour space is fully described by only nine components. This can be utilized to define the nine functions

$$f_0(x,p) = \mathrm{Tr}[F(x,p)] \quad , \quad f_a(x,p) = \mathrm{Tr}[\tfrac{1}{2}\lambda_a F(x,p)] \tag{10.11}$$

that can be identified with the classical averages $\int dQ\, f(x,p,Q)$ and $\int dQ\, Q^a\, f(x,p,Q)$. For these functions one can derive a set of coupled equations of motion from the Dirac equation:

$$p^\mu \partial f_0/\partial x^\mu = p_\mu F^{\mu\nu}_a \partial f_a/\partial p^\nu$$

$$p^\mu \partial f_a/\partial x^\mu - f_{abc} p^\mu A^b_\mu f_c - d_{abc} p_\mu F^{\mu\nu}_b (\partial f_a/\partial p^\nu) = 1/6\; p_\mu F^{\mu\nu}_a (\partial f_0/\partial p^\nu) \tag{10.12}$$

These equations are the analogues of eq.(6) derived from the quantum theory, neglecting contributions from the spin of the quarks[13].

In order to construct a full transport theory, the eqs.(12) must be supplemented by a distribution function for the gluons and by collision terms which connect one distribution function to itself but also to other functions. To describe the chemistry of the quark-gluon plasma, separate phase-space distributions for the various quark flavours should be introduced. In this way, a large number of coupled transport equations of the type (12) are obtained, which are probably much too complicated to be solved in the near future. A reasonable approach for an approximate solution would be to assume that deviations from the equilibrium state - in which the collision terms vanish - are small and to derive transport coefficients for the quark-gluon plasma from the linearized transport equations. These transport coefficients would describe the rate of approach to the equilibrium state for various reaction processes. Typical examples for such coefficients are thermal conductivity, colour conductivity [Ba79], viscosity, and quarko-chemical reaction rates, e.g. for the processes $q\bar{q}\to s\bar{s}$ or $gg\to s\bar{s}$.

The transport coefficients are then introduced into the momentum space-integrated transport equations to yield a coupled set of dissipative hydrodynamical equations. These relativistic Navier-Stokes equations, which describe the spatial evolution of the quark-gluon gas (or fluid) in interaction with an average colour field, show some promise of being solvable with available numerical techniques. In the absence of the terms arising from two-body collisions, the structure of these equations

---

[13] We mention that this approximation may not be really justified, because the interaction of the quark spin with the colour field is known to be important in some cases. E.g. it is responsible for the N-Δ mass splitting of 300 MeV [deG75].

is simple. There is one equation for the conservation of baryon current, one for the conservation of colour current, and one for energy-momentum conservation:

$$\partial_\mu b^\mu = 0 \quad , \qquad \partial_\mu (j_{Qa}^{\ \mu} + j_{Ga}^{\ \mu}) = - f_{abc} A_\mu^b (j_{Qc}^{\ \mu} + j_{Gc}^{\ \mu})$$

and

$$\partial_\nu (T_Q^{\mu\nu} + T_G^{\mu\nu}) = (j_{Q\nu}^{\ a} + j_{G\nu}^{\ a})\, F_a^{\mu\nu} \ . \tag{10.13}$$

Here $b^\mu(x)$ is the baryon current carried by the quark-antiquark fluid

$$b^\mu(x) \quad = \quad 1/3 \int (dp)\ \mathrm{Tr}[\gamma^\mu F(x,p)] \ , \tag{10.14}$$

while $j_{Qa}^{\ \mu}$ and $T_Q^{\mu\nu}$ for the quark fluid are given by the expressions (9) and (10). The contributions of the gluon fluid, $j_{Ga}^{\ \mu}$ and $T_G^{\mu\nu}$, must be derived in a similar way from the gluon phase-space distribution function, which is a 8×8 colour matrix. The gauge invariant separation of the gluonic degrees of freedom into an averaged, mean colour field and into the distribution function of real gluons is an interesting, but - to my knowledge - unsolved problem (see e.g. [Ro69] for a discussion of the same problem for a scalar field, or [BS59]).

The completion of this ambitious programme will require a lot of hard work to be done. However, even without full numerical solution of the transport equations, very interesting results can be obtained in two directions. One way of proceeding consists in taking eqs.(13) as a starting point and assuming that the average colour current vanishes. One is then left with the "standard" relativistic hydrodynamical equations [LL66] which must be solved with appropriate initial conditions for b^μ and $T^{\mu\nu}$. Work along this line will be discussed in the following chapter.

The other possible direction is to derive the magnitude of the transport coefficients from the linearized versions of the collision terms. This can provide valuable insight into the possible evolution of the quark-gluon plasma, because the transport coefficients allow to estimate the characteristic time-scales of reactions and various collective modes. Exploratory investigations into the possible effects of viscosity and heat conductivity of the plasma have recently been made by several authors [HK83,HST83,DG84,Ga85]. In the approximation where the particle interactions are parametrized by a fixed collision time τ, Gavin [Ga85] has found that the shear viscosity in the quark-gluon plasma can be written as

$$\eta = 4\tau\varepsilon/15 \qquad (10.15)$$

where ε is the energy density of quarks and gluons as given by eq. (3.7). The coefficients of bulk viscosity and heat conductivity are much smaller, mainly because the plasma is made up of essentially massless particles. Other interesting quantities which are relevant for the approach to equilibrium in the quark-gluon plasma are chemical reaction rates. In ch.12 we shall discuss one particularly interesting example, namely the production rate of strange quarks.

As we discussed in the previous chapter, the transport equations for the quark-gluon plasma can be reduced to relativistic hydrodynamical equations under suitable approximations. These equations state nothing else than the conservation of baryon number, energy, momentum and colour. If one assumes that the plasma is locally colour-neutral, the continuity equation for the colour current drops out, leaving us with the equations

$$\partial_\mu b^\mu = 0 \quad , \qquad \partial_\mu T^{\mu\nu} = 0 \ . \tag{11.1}$$

In order to obtain a simple solution to these equations, we further assume that the plasma behaves like an ideal gas, i.e. we neglect effects of viscosity and thermal conductivity. The energy-momentum tensor $T^{\mu\nu}$ can then be expressed by the local energy density ε, the pressure P and the local flow velocity $u^\mu(x,t)$ which is normalized to one:

$$u^\mu u_\mu \;=\; (u^0)^2 - (u)^2 \;=\; 1 \ . \tag{11.2}$$

In terms of the variables $\vec{\beta} = \vec{v}/c$ and $\gamma = (1-\beta^2)^{-\frac{1}{2}}$ the four-velocity u^μ can be written as $u^\mu = (\gamma, \vec{\beta}\gamma)$. The energy-momentum tensor then has the general form

$$T^{\mu\nu} \;=\; (\varepsilon+P)u^\mu u^\nu - Pg^{\mu\nu} \ , \tag{11.3}$$

where ε and P are functions of temperature T and quark chemical potential μ: $\varepsilon(\mu,T)$, $P(\mu,T)$. Similarly, the baryonic current b^μ can be expressed in the form

$$b^\mu \;=\; n_b u^\mu \ , \tag{11.4}$$

where $n_b(\mu,T)$ is the baryon number density in the local rest frame of the fluid. Including interactions perturbatively, ε, P and n_b are given in terms of μ and T by the expressions (8.7).

Because ε, P and n_b depend on μ and T, the equations for b^μ and $T^{\mu\nu}$ are coupled, in general. There is, however, one exception to this rule: if ε and P are - up to a factor - the same functions of μ and T, i.e.

$$P(\mu,T) = \alpha\,\varepsilon(\mu,T) \quad , \qquad \alpha \equiv (v_s/c)^2 \ , \tag{11.5}$$

P can be eliminated from $T^{\mu\nu}$ and ε can be taken as an independent variable in eq.(3). By studying small perturbations one can show that V_s in eq.(5) is the velocity of sound in the plasma. For an interacting gas of ultrarelativistic particles, i.e. particles with negligible rest mass, eq.(5) holds quite generally, with the value $\alpha = 1/3$. An example for this relation is provided by eqs.(8.7) for the equation of state of the perturbative quark-gluon plasma.

Using eqs. (5) and (3), the second equation in (1) now becomes

$$(1+\alpha)\partial_\mu(\varepsilon u^\mu u^\nu) - \alpha g^{\mu\nu}\partial_\mu\varepsilon = 0 \ . \tag{11.6}$$

By simple manipulations we can derive a very interesting equation from eq.(6). To do so, we contract eq.(6) with the velocity u_ν, and use the product rule of differentiation to evaluate the first term:

$$(1+\alpha)[u_\nu u^\nu(u^\mu\partial_\mu\varepsilon+\varepsilon\partial_\mu u^\mu) + \varepsilon u^\mu(u_\nu\partial_\mu u^\nu)] - \alpha u^\mu\partial_\mu\varepsilon = 0 \ . \tag{11.7}$$

Now eq.(2) implies that the last term in the square brackets vanishes, because

$$\partial_\mu(u_\nu u^\nu) = 2u_\nu\partial_\mu u^\nu = o \ . \tag{11.8}$$

Using eq.(2) once more in the first term in square brackets, eq.(7) reduces to

$$u^{\mu}\partial_{\mu}\varepsilon + (1+\alpha)\varepsilon\partial_{\mu}u^{\mu} = 0 . \tag{11.9}$$

Since u^{μ} is the derivative of the position x^{μ} of a fluid volume element with respect to its proper time, i.e. $cu^{\mu} = dx^{\mu}/d\tau$, the expression $u^{\mu}\partial_{\mu}$ is just

$$cu^{\mu}\partial_{\mu} = (dx^{\mu}/d\tau)\partial_{\mu} = d/d\tau . \tag{11.10}$$

Dividing by ε, eq.(9) now takes on a very simple form:

$$\varepsilon^{-1}(d\varepsilon/d\tau) = -(1+\alpha)c(\partial_{\mu}u^{\mu}) , \tag{11.11}$$

which is the equation we wanted to derive.

If $\partial_{\mu}u^{\mu}$ would be a function of the proper time only, eq.(11) could be integrated straightforwardly. Of course, this is usually not the case: $\partial_{\mu}u^{\mu}$ normally depends on all space-time variables and not only on $\tau = \sqrt{(t^2-x^2/c^2)}$. It is here that the inside-outside cascade discussed in ch.4 simplifies the picture. Remember that its basic assumption was that the products of a high-energy particle collision "materialize" only after a proper time τ_0, measured in the rest frame of the reaction product. At sufficiently high energy this formation time is highly dilated in the laboratory system, $t_0 = \gamma\tau_0$, i.e. the fragments materialize far downstream of the target. In other words, the reaction volume is strongly expanded in the longitudinal (beam) direction, if looked at in the laboratory (see Fig.37). In a first approximation it is, therefore, allowed to drop the transverse spatial dimensions and describe the reaction in one space (longitudinal) and one time dimension, x and t.

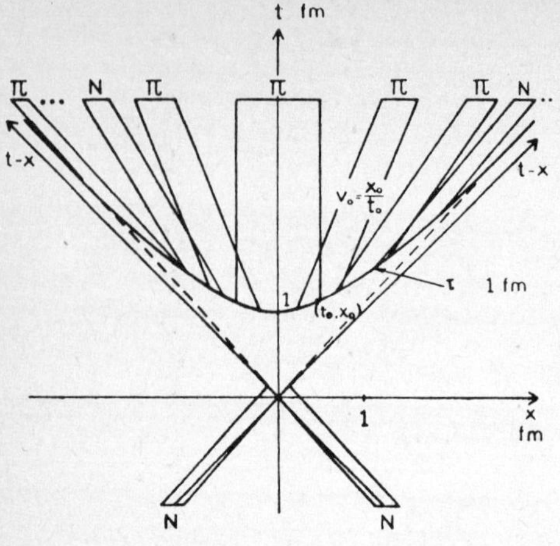

Fig. 37: Schematic space-time diagram of a collision between two nucleons at very high energy, as seen from the centre-of-mass frame. The nucleons collide at the origin and the produced particles appear at $\tau_0 \simeq 1$fm as real particles which can participate in further interactions (from [KRR83]).

Let us consider a particle produced at rapidity $y = \mathrm{artanh}(v/c)$, where v is the velocity in the longitudinal direction (see Fig.38). The particle moves on a straight line with the coordinate representation

$$(t,x) = (\tau/\gamma, v\tau/\gamma) . \tag{11.12}$$

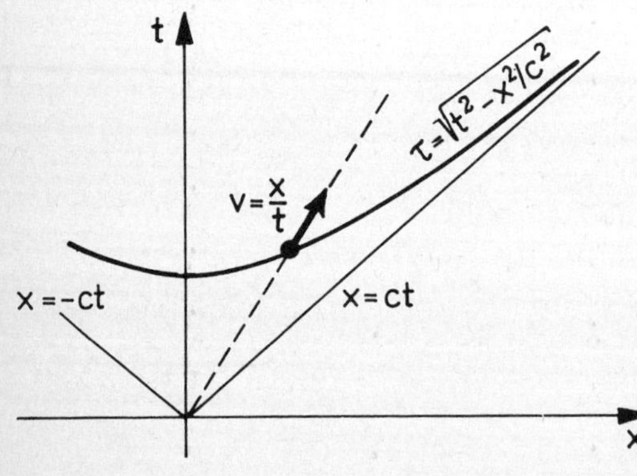

Fig. 38: Definition of the coordinates. The dashed line represents the world line of a particle produced at rapidity y.

The four-velocity of the particle is

$$u^\mu = (\gamma, \gamma v/c) = (t/\tau, x/c\tau) , \tag{11.13}$$

where we used eq.(12) to replace the Lorentz contraction factor γ by the coordinates. The calculation of the divergence of u^μ is now easy:

$$c\partial_\mu u^\mu = \partial(t/\tau)/\partial t + \partial(x/\tau)/\partial x$$

$$= 2/\tau - (t/\tau^2)\partial\tau/\partial t - (x/\tau^2)\partial\tau/\partial x = 1/\tau \quad , \tag{11.14}$$

where we made use of the relations $\partial\tau/\partial t = t/\tau$ and $\partial\tau/\partial x = x/(\tau c^2)$, which are easily proved by explicit differentiation.

As we have shown, $\partial_\mu u^\mu$ depends only on the proper time under the simplifying assumptions of the inside-outside cascade. If we further assume that the energy density ε of the created particles at the formation time τ_0 is everywhere the same, ε_0, independent of rapidity y, then eq.(11) can be integrated as an ordinary differential equation, starting from τ_0:

$$\ln(\varepsilon/\varepsilon_0) = -(1+\alpha)\ln(\tau/\tau_0) \quad ,$$

$$\varepsilon = \varepsilon_0(\tau_0/\tau)^{1+\alpha} = \varepsilon_0(\tau_0/\tau)^{4/3} \quad . \tag{11.15}$$

Here we put α=1/3. The scaling solution (15) was derived by Bjorken [Bj83] for the evolution of the energy density in the central region of a high-energy collision of two nuclei.[14] Of course, the assumption of a constant energy density ε_0 at the production time is crucial. There are good reasons to assume that this assumption is fairly well satisfied in the central rapidity region (see Fig.9). It certainly breaks down at the edges of the central region, when the target and projectile fragmentation regions are approached.

---

[14] Bjorken's solution is very similar to the solution obtained by Landau [La53] in the context of his hydrodynamical model of particle collisions at ultra-relativistic energies.

Kajantie, McLerran and their collaborators have calculated the initial energy density $\varepsilon(\tau_0,y)$ expected for a nucleus-nucleus collision in the framework of the inside-outside cascade model [KM82a,KM83,KRR83,GM84]. Even in one spatial dimension the hydrodynamical equations must then be integrated numerically. The numerical solution is found to approach the scaling solution (15) after some time, sooner in the central region ($y_{cm} \simeq 0$) and later in the fragmentation regions ($y_{cm} \simeq \pm\frac{1}{2}y_{lab}$). As a general result, the energy density falls off rapidly due to the expansion of the quark-gluon plasma fireball in the longitudinal direction. The transverse expansion is much slower [Ba83].

Due to the rapid longitudinal expansion the energy density is expected to stay only for a few fm/c above the critical density at which the deconfinement phase transition occurs. The question is then whether this time is sufficiently long to reach thermal equilibrium in the quark-gluon plasma, after the phase has been formed at time τ_0. This problem has been investigated by Ochiai et al. [Oc84]. When these authors required that the particles created at τ_0 have time to participate on the average in three further interactions (with 20 mb cross-section), they found that the energy density has fallen considerably when thermalisation is finally achieved. At 400 GeV per nucleon in the lab system they predict the thermalized energy density to fall below 1 GeV/fm^3 even for U+U collisions. If this is correct, colliding beam experiments are required to produce the quark-gluon plasma in the laboratory (see also [Ba84] for possible dynamical aspects of the phase transition).

During the past year, several authors have studied the possible effects of a non-ideal equation of state on the hydrodynamical evolution of the fireball. In particular, the influence of a finite viscosity was investigated [HK83,HST83,DG84,Ka84,Ga85]. Although there is some disagreement in technical details, these authors agree in their finding that viscosity slows the expansion of the fireball down, but not very much for realistic values of the viscosity (see Fig.39). Easily understandable is also that viscous flow leads to an increase in entropy. Still, further work is needed to obtain a full quantitative understanding of the influence of deviations from local equilibrium on the hydrodynamic expansion of the quark-gluon plasma.

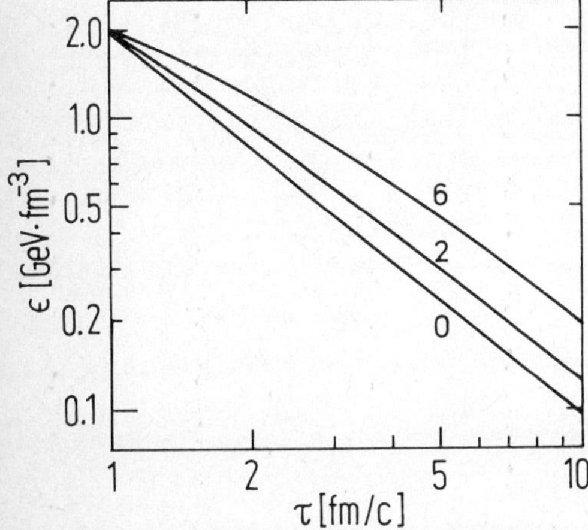

Fig. 39: Time-dependence of the energy density in the hydrodynamical model. The lines 0, 2, 6 correspond to three different choices of the viscosity parameter η/T^3 (from [DG84]).

12. STRANGENESS PRODUCTION

We have previously discussed that the amount of strangeness produced in a nuclear collision may be a signal for the transition to a quark-gluon phase. The argument then was based on the equilibrium density of strange quarks in the plasma. It still remains to be seen whether the plasma state lives sufficiently long, so that the equilibrium can be reached. In order to decide this question we must compare the expected lifetime of the plasma state with the chemical reaction rate for the production of strange quark-antiquark pairs in the quark-gluon plasma [RM82].

For the lifetime we shall take the very crude number $\tau_{QGP} \approx 6$ fm/c = 2×10^{-23}s. This guess is based on the assumption that the extension of the region containing nuclear matter in the plasma state is of the order of 4-6 fm, and the fact that the transition back into the hadronic phase cannot proceed with more than the speed of light. The quark-gluon plasma may live much longer, if unexpected events occur, but it can hardly exist for a shorter period of time only, if it is produced at all. This estimate is supported by recent schematic calculations done by Chin [Ch82], by Biró and Zimanyi [BZ83] and by Baym et al. [Ba83].

In order to calculate the reaction rate, we have to ask which elementary processes lead to the creation of strange quark pairs. In lowest order QCD four diagrams contribute:

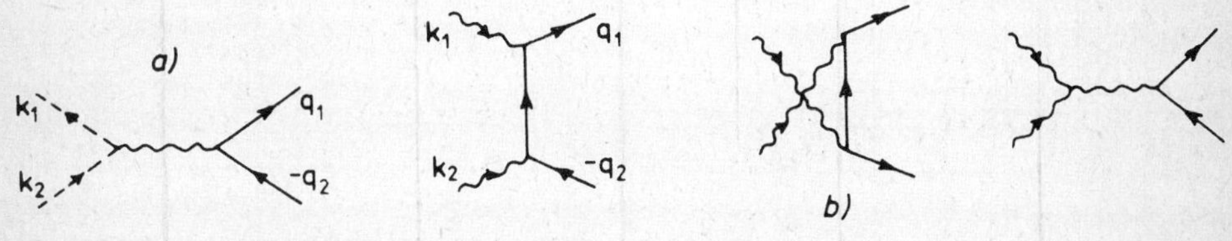

a)

b)

The diagram in part (a) of the figure denotes the annihilation of a pair of light quarks (u or d) into a virtual gluon which, in turn, is converted into a pair of strange quarks. This process is analogous to muon pair production in e^-e^+- collisions. The diagrams in part (b), which must be combined to maintain gauge-invariance, describe the conversion of a pair of real gluons into a strange quark pair. The four-momenta of the particles going into the reactions are denoted by k_1 and k_2, while those of the outgoing particles are p_1 and p_2. If the rest mass of the particles in the initial channel is neglected, the two invariant variables s (square of c.m. energy) and t (square of four-momentum transfer) are:

$$s = (k_1+k_2)^2 = 2(k_1k_2) = (p_1+p_2)^2 = 2m_s{}^2 + 2(p_1+p_2) \ ,$$

$$t = (k_1-p_1)^2 = m_s{}^2 - 2(k_1p_1) = m_s{}^2 - 2(k_2p_2) \ . \qquad (12.1)$$

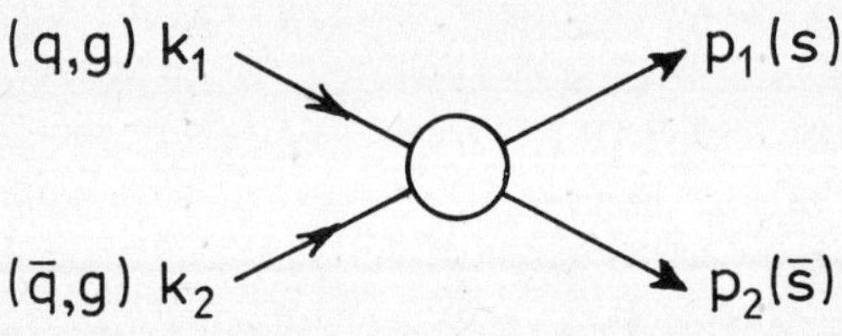

The amplitudes for those four diagrams contributing to strange particle production have been investigated by Combridge [Co79]. Because we are not interested in the quantum numbers (spin, colour) of the particles in the final state, and since we shall assume statistical distributions for the particles in the initial state, the squared amplitudes must be *summed* over final state colours and spins and *averaged* over the colour and spin quantum numbers in the initial state. The averaged transition probability thus obtained is still a function of both variables s and t. Since we are interested only in the total reaction rate, it is permitted to integrate over the momenta of the strange quarks in the final state. In this way we get rid of the variable t and obtain for the averaged cross-section:

$$\sigma = 1/16 \int(dp_1)\int(dp_2)\ (k_1k_2)^{-1}\ |M|^2\ (2\pi)^4\delta(p_1+p_2-k_1-k_2) =$$
$$= 1/16\pi s^2 \int dt\ |M|^2 .$$
(12.2)

The lowest-order QCD diagrams shown above yield the following expressions for the averaged cross-sections as function of total c.m. energy (squared) s:

$$\sigma_{q\bar{q}\to s\bar{s}} = 8\pi\alpha_s{}^2/27s\ (1+2m^2/s)w(s)$$
$$\sigma_{gg\to s\bar{s}} = 2\pi\alpha_s{}^2/3s\ [(1+4m^2/s+m^4/s^2)\tanh^{-1}w(s) - (7/8+31m^2/8s)w(s)] ,$$
(12.3)

where $w(s) = \sqrt{(1-m^2/s)}$. In the amplitude M the coupling constant α_s is originally a function of both s and t. Due to the integration over t in eq.(2) the value of α_s to be used in eq.(3) is a weighted average over space- and time-like transferred momenta. Therefore, it is not quite clear what the appropriate value of α_s in eq.(3) is. Because the production of strange quarks takes place at typical momenta of 300-500 MeV/c, we take the value

$$\alpha_s = 0.6 \qquad \text{in connection with} \qquad m_s = 150\ \text{MeV}.$$
(12.4)

This is simply a guess which is not motivated by more than being an interpolation of the MIT-bag model fit parameters ($\alpha_s \simeq 2.2$) and the results from high-energy e^-e^+-scattering ($\alpha_s \simeq 0.2$).

To obtain the strange quark-pair production rate per time and unit volume, the averaged cross-sections must be integrated over the momentum space distributions of light quarks, antiquarks and gluons in the initial state. In order to avoid introducing additional parameters we make the simplest possible assumption[15], i.e. that light quarks and gluons are in thermal equilibrium determined by a temperature T. For the light quarks we also have to specify a chemical potential μ to fix the baryon number density in the quark-gluon plasma. Under these assumptions the distribution functions are

$$\rho_g = (e^{\beta p}-1)^{-1}, \qquad \rho_q = (e^{\beta(p-\mu)}+1)^{-1}, \qquad \rho_{\bar{q}} = (e^{\beta(p+\mu)}+1)^{-1}, \tag{12.5}$$

where we have assumed gluons and light quarks to be massless. With these distributions the reaction rate for strange quark-pair production is given by

$$A = dN_s/d^3xdt \;=\; \int(dk_1)\int(dk_2)\int ds^2\ \delta[s-(k_1+k_2)^2]\times$$
$$\times\ [(2\times 8)^2\rho_g(k_1)\rho_g(k_2)\sigma_{gg\to s\bar{s}}(s) + 2\times(2\times 3)^2\rho_q(k_1)\rho_{\bar{q}}(k_2)\sigma_{q\bar{q}\to s\bar{s}}(s)] \tag{12.6}$$

where the numerical factors count spin, colour and flavour degrees of freedom.

As more and more strange quarks are produced and their concentration approaches the value taken in chemical equilibrium, eq.(3), the reverse reaction will become increasingly important. At equilibrium concentration the reverse reaction rate precisely balances the forward reaction rate, and the net rate of production of strange quarks vanishes. If we can neglect the effect of the Pauli principle for strange quarks, as we did in the considerations leading to eq.(3), the forward reaction rate is independent of the strangeness concentration. If we further make the assumption[16] that the already produced strange quarks and antiquarks move

---

[15] This certainly leads to a conservative estimate. If the momentum distribution of quarks and gluons is anisotropic, it will be easier to overcome the threshold $2m_s^2$ for the particles moving in the preferred momentum direction. Biró and Zimanyi [BZ83] have found that the anisotropy must be very large, of order 1:10, to produce a sizable effect.

freely and independently through the plasma, the reverse reaction rate must grow as the square of the concentration of strange quarks $n_s = dN_s/d^3x$ (because $N_s = N_{\bar{s}}$). We can then write the following rate equation for the approach to chemical equilibrium of the strangeness concentration:

$$dn_s/dt = A[1-(n_s(t)/n_s^{eq})^2] \ . \tag{12.7}$$

The solution of this equation

$$n_s(t) \ = \ n_s^{eq} \tanh(At/n_s^{eq}) \ \rightarrow \ n_s^{eq} \ (1-2e^{-2t/\tau}) \tag{12.8}$$

introduces a characteristic relaxation time

$$\tau = n_s^{eq}/A \tag{12.9}$$

for the approach to chemical equilibrium.

While n_s^{eq} is given by eq.(3), the basic rate A must be obtained by numerical integration of eq.(6). Taking the standard value μ = 300 MeV for the light quark chemical potential,one finds that τ is of the order of 10^{-23}s, decreasing with growing temperature (see Fig.42). The figure compares the total reaction rate with the rate for the process $q\bar{q}\rightarrow s\bar{s}$ alone, which was investigated by Biró and Zimanyi [BZ82]. One sees, somewhat surprisingly, that the rate of the process involving only quarks is an order of magnitude smaller than the total production rate, which means that the gluonic creation mechanism is by far the dominant process. The gluonic contribution to eq.(6) can be evaluated exactly by expanding the Bose distributions into power series in exp(-k/T). This gives the result

---

[16] Whether this assumption is justified depends on the transport properties of the plasma, e.g. its viscosity, and on the strength and range of the interaction between the quarks.

$$\tau \simeq \tau_g = 9/7\ \alpha_s^{-2}\ (2\pi m_s T^3)^{-\frac{1}{2}}\ \exp(m_s T)\ (1+99T/56m_s+\ldots)^{-1}\ , \tag{12.10}$$

which clearly exhibits the functional dependence of the reaction time on the parameters T, m_s, α_s.

This reaction time-scale must be compared with the expected lifetime of the quark-gluon plasma state which we estimated to be 2×10^{-23}s. Fig.42 shows that the strangeness production time τ is within this limit for plasma temperatures above 160 MeV. In order to find out how close the plasma comes to chemical equilibrium with respect to strangeness, one has to plot the time-evolution of the strangeness concentration $n_s(t)$, given by eq.(8). This is done in Fig.43 which shows that even for fairly "low" temperatures, such as 120 MeV, the concentration of strange quarks reaches half its equilibrium value during the estimated lifetime of the quark-gluon plasma.

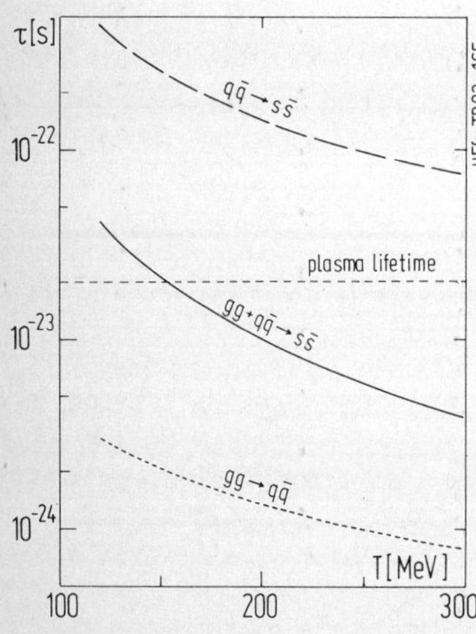

Fig. 42: Characteristic equilibration time for strangeness abundance in the quark-gluon plasma. The gluonic production mechanism is dominant. The lowest curve gives an indication of the thermalization time for the plasma phase.

At this point, the reader may wonder whether other heavier quark flavours, e.g. charm or beauty, would not yield even better signatures of the plasma

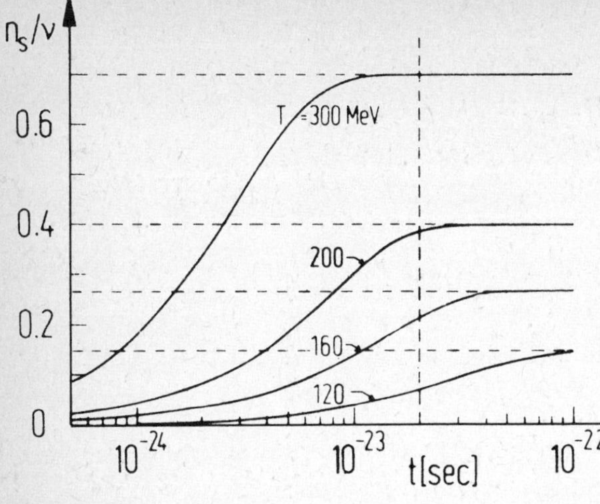

Fig. 43: Strangeness per baryon number as function of time for various temperatures of the quark-gluon plasma.

state than strangeness. One can estimate the production rate of charmed quarks along the same line, as done by Cleymans et al. [CP84,CV84], but it is uncertain whether the result has any meaning. The uncertainty originates in the momentum space distribution functions of the reactants, eq.(5), where we assumed thermal equilibrium. It is well known, however, that more and more collisions among the particles of a gas are required to bring the distribution into equilibrium shape as one goes away from the peak of the distribution into the tail region. In order to produce a charmed quark pair, the combined energy of the particles in the initial state must exceed twice the charmed quark mass $m_c \simeq 1500$ MeV. This is about ten times the thermal energy per degree of freedom (for $T = 160$ MeV), and thus it is extremely doubtful whether there is time enough to populate this region in the tail of the momentum distribution.

In addition, the finite number of particles contained in the plasma will lead to a suppression of the high-momentum components as compared with the canonical distribution. Since even eq.(10) predicts a long reaction time for charmed particle production due to the factor $\exp(m_c/T)$, we will content ourselves in saying that charmed particles may not constitute such a good signal for the quark-gluon plasma. Finally, the possible use of exotic meson states as a signal has been discussed by Liu [Li83].

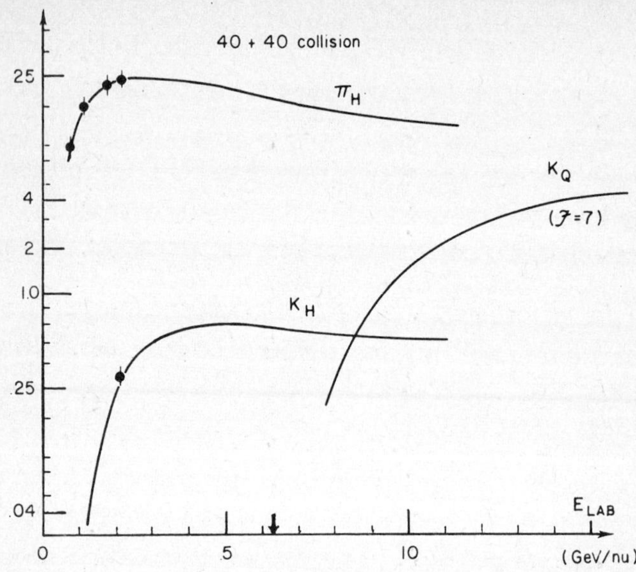

Fig. 44: Number of produced pions and kaons in Ar-Ca collisions versus bombarding energy. "H" labels the hadronic phase, "Q" the plasma phase. The phase transition is found to occur at the position of the arrow within the model used to describe the hydrodynamics. Observe the rapid increase of the kaon yield when the quark-gluon phase is reached (from [BZ83]).

Finally, we turn our attention to the dotted curve in Fig. 42, which represents the reaction time for the process $gg \to q\bar{q}$, i.e. the conversion of gluons into light quarks. Here a mass parameter of 15 MeV was used for the light quarks. The reader must be cautioned that this result contains some uncertainties, because the cross-section $\sigma_{gg \to q\bar{q}}$ - identical in form with the second expression (3) - diverges in the limit $s \to 0$. The strong coupling constant α_s is also not well known at such small energies. Whatever these uncertainties may be, the dotted line in Fig.44 shows that the reaction $gg \to q\bar{q}$ and its reverse are extremely fast processes, with a characteristic time of or below 1 fm/c = 3×10^{-24} s. This result bears importance for the formation phase of the quark-gluon plasma, because it tells us that gluons and light quarks will reach mutual chemical equilibrium very soon, no matter what their relative abundance has been in the prior hadronic phase. In this respect we can feel reassured of our basic assumptions about the nature of the quark-gluon plasma.

We conclude this section by stating once more the primary result: Strangeness is produced fast enough in the quark-gluon plasma to come close to the prediction of equilibrium thermodynamics. This remains true if one takes account of the expansion and consequent cooling of the hot plasma fireball. For a simple model involving isotropic self-similar expansion Biró and Zimányi [BZ83] found that strange particle production grows rapidly after the plasma phase is reached (see curve labelled "K_Q" in Fig.44) and even comes close to the predicted yield of pions out of the hadronic phase [Bi82,Ba83a]. As we pointed out, this spectacular increase is essentially due to the abundant presence of gluons in the plasma phase. Strange particles are a signal for gluons.

REFERENCES

General surveys of the field are found in the proceedings of the "Quark Matter" conferences: Bielefeld (1982) in [JS82], Brookhaven (1983) in Nucl. Phys. A418 (1984), and Helsinki (1984) to be published.

Ad81 S.Adler, Phys.Rev. D23 (1981) 2905

AGD63 A.A.Abrikosov, L.P.Gorkov, and I.E.Dzyaloshinsky, Methods of Quantum Field Theory in Statistical Physics, Prentice-Hall, Englewood Cliffs 1963

AKM80 R.Anishetty, P.Köhler and L.McLerran, Phys.Rev. D22 (1980)2793

AL73 E.S.Abers and B.Lee, Phys.Reports 9 (1973) 1

Al84 M.Althoff et al., TASSO collab., Phys.Lett. 139B (1984) 126

AO80 J.Ambjorn and P.Olesen, Nucl.Phys. B170 (1980) 60

Ar82 H.Arodz, Phys.Lett. 116B (1982) 251

Ba79 G.Baym, Physica 96A (1979) 131

Ba83 G.Baym, B.L.Friman, J.P.Blaizot, M.Soyeur, and W.Czyz, Nucl.Phys. A407 (1983) 541

Ba83a H.W.Barz, T.S.Biró, B.Lukács, and J.Zimányi, Z.Phys. A311 (1983) 311

Ba84 H.W.Barz, B.Kämpfer, L.P.Csernai, and B.Lukács, Phys.Lett. 143B (1984) 334

BBZ81 M.Baker, J.S.Ball, and F.Zachariasen, Nucl.Phys. B186 (1981) 531 + 560

BC78 K.M.Bitar and S.J.Chang, Phys.Rev. D17 (1978) 486 and D18 (1978) 435

Be74 C.Bernard, Phys.Rev. D9 (1974) 3312

BG84 W.Busza and A.S.Goldhaber, Phys.Lett. 139B (1984) 235

BGM83 B.Banerjee, N.K.Glendenning, and T.Matsui, Phys.Lett. 127B (1983) 453

Bi82 T.S.Biró, B.Lukács, J.Zimányi, and H.W.Barz, Nucl.Phys. A386 (1982) 671

BI83 J.Baacke and Y.Igarashi, Phys.Rev. D27 (1983) 460

Bj83 J.D.Bjorken, Phys.Rev. D27 (1983) 140

BM69 L.S.Brown and G.J.Maclay, Phys.Rev.184 (1969) 1272

BMS77 I.A.Batalin, S.G.Matinyan, and G.K.Savvidi, Sov.J.Nucl.Phys. 26 (1977) 214

Bo67 P.N.Bogolioubov, Ann.Inst.Henri Poincaré 8 (1967) 163

BR79 G.E.Brown and M.Rho, Phys.Lett. 82B (1979) 177

Br81 R.Brandelik et al., TASSO collab., Phys.Lett. 100B (1981) 357

BRV79 G.E.Brown, M.Rho, and V.Vento, Phys.Lett. 84B (1979) 83

BS59 N.N.Bogoliubov and D.V.Shirkov, Introduction to the Theory of Quantized Fields, Wiley, New York 1959

BS81 R.Bock and R.Stock (Eds.), Workshop on Future Relativistic Heavy Ion Experiments, GSI-Report 81-6

BU83 T.Banks and A.Ukawa, Nucl.Phys. B225 (1983) 145

Bu83 T.H.Burnett et al., JACEE collab., Phys.Rev.Lett. 50 (1983) 2062 and 51 (1983) 1010

Bu84 T.H.Burnett et al., JACEE collab., preprint, Tokyo 1984

BZ82 T.Biró and J.Zimanyi, Phys.Lett. 113B (1982) 6

BZ83 T.Biró and J.Zimanyi, Nucl.Phys. A395 (1983) 525

Ca74 P.Capiluppi et al., Nucl.Phys. B70 (1974) 1

CCR84 J.W.Clark, J.Cleymans, and J.Rafelski, preprint UCT-TP-18-2, Cape Town 1984

CDH83 J.Cleymans, M.Dechantsreiter and F.Halzen, Z.Phys. C17 (1983) 341

CES83 T.Celik, J.Engels, and H.Satz, Phys.Lett. 129B (1983) 323

Ch74 A.Chodos, R.L.Jaffe, K.Johnson, C.B.Thorn, and V.F.Weisskopf, Phys.Rev. D9 (1974) 3471

Ch78 S.A.Chin, Phys.Lett. 78B (1978) 552

Ch82 S.A.Chin, Phys.Lett. 119B (1982) 51

CK84 L.P.Csernai and J.Kapusta, Phys.Rev.D29 (1984) 2664

CKS81 A.Cabo, O.K.Kalashnikov and A.Shabad, Nucl.Phys. B185 (1981) 473 3

CNN79 A.Casher, H.Neuberger, and S.Nussinov, Phys.Rev. D20 (1979) 179

Co79 B.L.Combridge, Nucl.Phys. B151 (1979) 429

CP84 J.Cleymans and R.Philippe, Z.Phys. C22 (1984) 271

Cr80 M.Creutz, Phys.Rev. D21 (1980) 2308

CS82 M.Crawford and D.Schramm, Nature 298 (1982) 538

CV84 J.Cleymans and C.Vanderzande, Phys.Lett. 147B (1984) 186

CW73 S.Coleman and E.Weinberg, Phys.Rev. D7 (1973) 1888

CZ83 P.Carruthers and F.Zachariasen, Rev.Mod.Phys. 55 (1983) 245

deG75 T.deGrand, R.L.Jaffe, K.Johnson, and J.Kiskis,Phys.Rev. D12 (1975) 2060

DG81 G.Domokos and J.I.Goldman, Phys.Rev. D23 (1981) 203

DG85 P.Danielewicz and M.Gyulassy, Phys.Rev. D31 (1985) 53

DGS85 S.Date, M.Gyulassy, and H.Sumiyoshi, "Nuclear Stopping Power at High Energy", preprint, Tokyo 1985

DJ80 J.F.Donoghue and K.Johnson, Phys.Rev. D21 (1980) 1975

DK84 T.DeGrand and K.Kajantie, Phys.Lett. 147B (1984) 273

Do83 G.Domokos, Phys.Rev D28 (1983) 123

DR83 M.Danos and J.Rafelski, Phys.Rev.D27 (1983) 671

DS81 W.Dittrich and V.Schanbacher, Phys.Lett. 100B (1981) 415

Du76 A.Duncan, Phys.Rev. D13 (1976) 2866

EGR80 T.Elze, W.Greiner, and J.Rafelski, J.Phys. G6 (1980) L419

EGR83 T.Elze, W.Greiner, and J.Rafelski, Phys.Lett. 124B (1983) 515

EGR84 T.Elze, W.Greiner, and J.Rafelski, Z.Phys. C24 (1984) 361

EKS82 J.Engels, F.Karsch, and H.Satz, Phys.Lett. 113B (1982) 398

ELM83 H.Ehtamo, J.Lindfors, and L.McLerran, Z.Physik C18 (1983) 341

En81 J.Engels, F.Karsch, I.Montvay, and H.Satz, Phys.Lett. 101B (1981) 89

En81a J.Engels, F.Karsch, I.Montvay, and H.Satz, Phys.Lett. 102B (1981) 332

En82 J.Engels, F.Karsch, H.Satz, and I.Montvay, Nucl.Phys. B205 (1982) 545

Fa84 M.A.Faessler, CERN preprint EP/84-78, June 1984

Fe76 E.L.Feinberg, Nuovo Cimento 34A (1976) 391

FF78 R.D.Field and R.P.Feynman, Nucl.Phys. B136 (1978) 1

FG72 H.Fritzsch and M.Gell-Mann, Proceed. 16th Intern. Conf. on

High-Energy Physics, Chicago 1972, vol.2

FL77 R.Friedberg and T.D.Lee, Phys.Rev. D15 (1977) 1694 and D16 1096

FL78 R.Friedberg and T.D.Lee, Phys.Rev. D18 (1978) 2623

Fl80 H.Flyvvbjerg, Nucl.Phys. B176 (1980) 379

FP67 L.D.Faddeev and V.N.Popov, Phys.Lett. 25B (1967) 291

FRS84 F.Fucito, C.Rebbi, and S.Solomon, Nucl.Phys. B248 (1984) 615

Fu80 R.Fukuda, Phys.Rev. D21 (1980) 485

FW71 A.L.Fetter and J.D.Walecka, Quantum Theory of Many-Particle Systems, McGraw-Hill, New York 1971

Ga85 S.Gavin, Nucl.Phys. A435 (1985) 826

Ge62 M.Gell-Mann, Phys.Rev. 125 (1962) 1067

GG61 S.L.Glashow and M.Gell-Mann, Ann.Phys. 15 (1961) 437

Gi74 R.Gilmore, Lie Groups, Lie Algebras and some of their Applications, Wiley, New York 1974

GL60 M.Gell-Mann and M.Lévy, Nuovo Cimento 16 (1960) 705

GLP84 R.V.Gavai, M.Lev, and B.Petersson, Phys.Lett. 149B (1984) 492

GLW80 S.R.de Groot, W.A.van Leeuwen, and C.G.van Weert, Relativistic Kinetic Theory, Amsterdam 1980

GM83 N.K.Glendenning and T.Matsui, Phys.Rev. D28 (1983) 2890

GM84 M.Gyulassy and T.Matsui, Phys.Rev. D29 (1984) 419

Go78 A.S.Goldhaber, Nature 275 (1978) 114

Go82 J.Gonzales, Nucl.Phys. B204 (1982) 485

Go83 M.I.Gorenstein, S.I.Lipskikh, V.K.Petrov, and G.M.Zinovjev, Phys.Lett. 123B (1983) 437

GR84 N.K.Glendenning and J.Rafelski, preprint LBL-17938, Berkeley 1984

Gu81 A.Guth, Phys.Rev. D23 (1981) 437

GW73 D.J.Gross and F.Wilczek, Phys.Rev.Lett. 30 (1973) 1343, Phys.Rev. D8 (1973) 3633

Gy83a M.Gyulassy, in: High Energy Nuclear Physics, Proc. 6th Balaton Conference, ed. J.Erö, Budapest 1983, p. 489

Gy83b M.Gyulassy, Nucl.Phys. A400 (1983) 31c

Gy84 M.Gyulassy, K.Kajantie, H.Kurki-Suonio, and L.McLerran, Nucl.Phys. B237 (1984) 477

Gy84a M.Gyulassy, Nucl.Phys. A418 (1984) 59c

Ha81 P.Hasenfratz, R.R.Horgan, J.Kuti, and J.M.Richard, Physica Scripta 23 (1981) 917

He82 U.Heinz, in [JS82], p. 439

He83 U.Heinz, Phys.Rev.Lett. 51 (1983) 351

HK78 P.Hasenfratz and J.Kuti, Phys.Reports 40 (1978) 75

HK83 A.Hosoya and K.Kajantie, preprint HU-TFT-83-62, Helsinki 1983

Ho82 C.Hogan, Phys.Lett. 133B (1982) 172

HS76 K.Huang and D.R.Stump, Phys.Rev. D14 (1976) 223

HSG84 U.Heinz, P.R.Subramanian, and W.Greiner, Z.Phys. A318 (1984) 247

HST83 A.Hosoya, M.Sakagami, and M.Takao, preprint OU-HET-53, Osaka 1983

Hw84 R.C.Hwa, Phys.Rev.Lett. 52 (1984) 492

Jo75 K.Johnson, Acta Phys. Pol. B6 (1975) 865

JS82 M.Jacob and H.Satz (Eds.), Quark Matter Formation and Heavy Ion Collisions, World Scientific Publ., Singapore, 1982

Ka79 J.I.Kapusta, Nucl.Phys. B148 (1979) 461

Ka79a J.I.Kapusta, Phys.Rev. D20 (1979) 989

Ka81 J.I.Kapusta, Nucl.Phys. B190 (1981) 425

Ka82 K.Kajantie, in [JS82], p. 39

Ka82a J.I.Kapusta, in [JS82], p. 115

Ka83 O.K.Kalashnikov, QCD at Finite Temperature, preprint, Moscow 1983

Ka83a C.G.Källman, Phys.Lett. 126B (1983) 366

Ka84 J.Kapusta, Phys.Lett. 136B (1984) 201

Ka84a K.Kajantie, Nucl.Phys. B418 (1984) 41c

Ke84 A.D.Kennedy, J.Kuti, S.Meyer, and B.J.Pendleton, preprint NSF-ITP-84-61, Santa Barbara 1984

KE83 G.Kälbermann and J.M.Eisenberg, Phys.Rev. C28 (1983) 1318

Ki80 L.S.Kisslinger, Direct photon production from quark matter in nuclear collisions, preprint, Pittsburgh 1980

KK79 O.K.Kalashnikov and V.V.Klimov, Phys.Lett. 88B (1979) 328

KK80 O.K.Kalashnikov and V.V.Klimov, Sov.J.Nucl.Phys. 31 (1980) 699

KK82 K.Kajantie and J.Kapusta, Phys.Lett. 110B (1982) 229

KK84 K.Kajantie and J.Kapusta, Ann.Phys.(NY) in print

KM77 M.B.Kislinger and P.D.Morley, Phys.Lett.67B (1977) 371

KM81 K.Kajantie and H.I.Miettinen, Z.Physik C9 (1981) 341

KM82 K.Kajantie and H.I.Miettinen, Z.Physik C14 (1982) 357

KM82a K.Kajantie and L.McLerran, Phys.Lett. 119B (1982) 203

KM83 K.Kajantie and L.McLerran, Nucl.Phys. B214 (1983) 261

KP83 J.Kuti and J.Polónyi, Some applications of a new stochastic method in lattice theories, Santa Barbara preprint NSF-ITP-83-108

KPS81 J.Kuti, J.Polónyi, and K.Szlachányi, Phys.Lett. 98B (1981) 199

KR83 K.Kajantie and R.Raitio, Phys.Lett. 121B (1983) 415

KR85 P.Koch and J.Rafelski, private communication, 1985

KRR83 K.Kajantie, R.Raito, and P.V.Ruuskanen, Nucl.Phys. B222 (1983) 152

KT82 E.Kolb and M.Turner, Phys.Lett. 115B (1982) 99

KT83 C.G.Källman and T.Toimela, Phys.Lett. 122B (1983) 409

La53 L.D.Landau, Izv.Akad.Nauk SSSR 17 (1953) 51

LD83 J.Lodenquai and V.Dixit, Phys.Lett. 124B (1983) 317

Le67 M.Lévy, Nuovo Cimento 52A (1967) 23

Le79 T.D.Lee, Phys.Rev. D19 (1979) 1802

Li83 K.F.Liu, preprint UK-83-06, Univ.of Kentucky 1983

LL66 L.D.Landau and E.M.Lifschitz, Lehrbuch der Theoretischen Physik 6 (Hydrodynamik), Akademie-Verlag, Berlin 1966

LM83 S.P.Li and L.McLerran, Nucl.Phys. B214 (1983) 417

LP79 P.Langacker and H.Pagels, Phys.Rev. D19 (1979) 2070

LPS84 J.A.Lopez, J.C.Parikh, and P.J.Siemens, preprint, Texas A+M, 1984

McL82 L.McLerran, in [JS82], p. 63

Me68 J.Mehra, Physica 37 (1968) 145

ME85 B.Müller and J.M.Eisenberg, Nucl.Phys. A435 (1985) 791

Mi83 K.Milton, Phys.Rev. D27 (1983) 439

MK79 P.Morely and L.Kislinger, Phys.Reports 51 (1979) 63

Mo83 J.Morishita et al., Z.Physik C19 (1983) 167

MP78 W.Marciano and H.Pagels, Phys.Reports 36 (1978) 137

MP82 I.Montvay and E.Pietarinen, Phys.Lett. 110B (1982) 148 and 115B (1982) 151

MR81 B.Müller and J.Rafelski, Phys.Lett. 101B (1981) 111

MS81 L.McLerran and B.Svetitsky, Phys.Lett. 98B (1981) 195 and Phys.Rev. D24 (1981) 450

Na82 S.Nadkarni, Phys.Rev. D27 (1983) 917

Ni81 N.N.Nikolaev, Sov.J.Part.Nucl. 12 (1981) 63

NN79 H.B.Nielsen and M.Ninomiya, Nucl.Phys. B156 (1979) 1

NO78 N.K.Nielsen and P.Olesen, Nucl.Phys. B144 (1978) 376

NO78a N.K.Nielsen and P.Olesen, Phys.Lett. 79B (1978) 304

NO79 H.B.Nielsen and P.Olesen, Nucl.Phys. B160 (1979) 380

NS81 M.Ninomiya and N.Sakai, Nucl.Phys. B190 (1981) 316

Oc84 T.Ochiai, S.Date, N.Suzuki, O.Miyamura, and H.Sumiyoshi, Phys.Lett. 142B (1984) 69

O181 K.Olive, Nucl.Phys. B190 (1981) 483

PDG82 Particle Data Group, Phys.Lett. 111B (1982)

PHJ82 C.Peterson, T.H.Hanson, and K.Johnson, Phys.Rev. D26 (1982) 415

Po84 J.Polonyi, H.W.Wyld, J.B.Kogut, J.Shigemitsu, and D.K.Sinclair, Phys.Rev.Lett. 53 (1984) 644

PT78 H.Pagels and E.Tomboulis, Nucl.Phys. B143 (1978) 485

PW84 R.Pisarski and F.Wilczek, Phys.Rev. D29 (1984) 338

Ra76 J.Rafelski, Phys.Rev. D14 (1976) 2358

Ra77 J.Rafelski, Phys.Rev. D16 (1977) 1890

Ra81 J.Rafelski, in [BS81], p. 282

Ra82 J.Rafelski, Phys.Reports 88 (1982) 331

Ra84 J.Rafelski, Nucl. Phys. A418 (1984) 489c

RD83 J.Rafelski and M.Danos, Perspectives in High-Energy Nuclear Collisions, preprint NBSIR 83-2725, Washington 1983

RM76 J.Rafelski and B.Müller, Phys.Rev. D14 (1976) 3532

RM82 J.Rafelski and B.Müller, Phys.Rev.Lett. 48 (1982) 1066

Ro69 P.Roman, Introduction to Quantum Field Theory, ch. 5.1, Wiley, New York 1969

RT80 K.Redlich and L.Turko, Z.Physik C5 (1980) 201.

Sa77 G.K.Savvidy, Phys.Lett. 71B (1977) 133

Sa82 H.Satz, New States of Matter, in: Proc.Int.Conf. on Nucleus-Nucleus Collisions, East Lansing, Sept. 1982

Sch58 J.Schwinger, Ed., Quantum Electrodynamics, Dover, New York 1958, papers no. 8 (V.S.Weisskopf) and 20 (J.Schwinger)

SF83 B.Svetitsky and F.Fucito, Phys.Lett. 131B (1983) 165

SGB81 H.Stöcker, M.Gyulassy, and J.Boguta, Phys.Lett. 103B (1981) 269

Sh78 E.V.Shuryak, Sov.Phys.JETP 47 (1978) 212

Sh80 E.V.Shuryak, Phys.Reports 61 (1980) 71

Sh83 E.V.Shuryak, Non-perturbative Phenomena in QCD Vacuum, Hadrons and Quark-Gluon Plasma, preprint CERN-83-01

Sh84 E.V.Shuryak, Theory and Phenomenology of the QCD Vacuum (1-8) 8 preprints, Novosibirsk 1984

SO84 D.Schramm and K.Olive, Nucl.Phys. A418 (1984) 289c

SR84 A.Schnabel and J.Rafelski, Diploma thesis, Frankfurt 1984

St81 P.M.Stevenson, Phys.Rev. D23 (1981) 2916

St82 R.Stock et al., proposal PS-190, in [JS82], p. 557

St84 H.Stöcker, Nucl.Phys. A418 (1984) 587c

St84a J.D.Stack, Phys.Rev. D29 (1984) 1214

St85 J.Staadt, Diploma thesis, to be published, Frankfurt 1985

Su77 L.Susskind, Phys.Rev D16 (1977) 3031

Su82 E.Suhonen, Phys.Lett. 119B (1982) 81

SY81 H.Sato and K.Yazaki, Phys.Lett. 98B (1981) 153

SZ80 E.V.Shuryak and O.V.Zhirov, Phys.Lett. 89B (1980) 253

tH71 G.t'Hooft, Nucl.Phys. B33 (1971) 173 and B35 (1971) 167

Th82 A.W.Thomas, Chiral symmetry and the bag model, preprint TH.3368-CERN and TRI-PP-82-29 (1983)

TMT80 S.Théberge, A.W.Thomas, and G.A.Miller, Phys.Rev. D22 (1980) 2838

To82 T.Toimela, Helsinki preprint HU-TFT-82-37

To83 T.Toimela, Z.Physik C17 (1983) 365

TTM81 A.W.Thomas, S.Théberge, and G.A.Miller, Phys.Rev. D24 (1981) 216

Tu81 L.Turko, Phys.Lett. 104B (1981) 153

Ut56 R.Utiyama, Phys.Rev. 101 (1956) 1597

Va83 D.Vasak, K.H.Wietschorke, B.Müller, and W.Greiner, Z.Physik C21 (1983) 119

vH82 L.van Hove, Phys.Lett. 118B (1982) 138

vH83 L.van Hove, Z.Phys. C21 (1983) 93

vH84 L.van Hove, CERN preprint TH.4055/84 (1984)

vH85 L.van Hove, CERN preprint TH.3924 (Z.Phys. C in print 1985)

Vi75 P.Vinciarelli, Nucl.Phys. B89 (1975) 463

VR80 R.D.Viollier and J.Rafelski, Helv.Phys.Acta 53 (1980) 352

VT65 V.S.Vanyashin and M.V.Terent'ev, Sov.Phys.JETP 21 (1965) 375

We72 S.Weinberg, Gravitation and Cosmology, Wiley, New York 1972

We73 S.Weinberg, Phys.Rev.Lett. 31 (1973) 494

Wi74 K.Wilson, Phys.Rev. D10 (1974) 2445

Wi84 E.Witten, Phys.Rev. D30 (1984) 272

Wo70 S.K.Wong, Nuovo Cimento 65A (1970) 689

Wo84 C.Y.Wong, Phys.Rev.Lett. 52 (1984) 1393 and Phys.Rev. D30 (1984) 961 + 972

YM54 C.N.Yang and R.L.Mills, Phys.Rev. 96 (1954) 191

Yo82 G.B.Yodh, in [JS82], p. 213

Lecture Notes in Physics

Vol. 195: Trends and Applications of Pure Mathematics to Mechanics. Proceedings, 1983. Edited by P.G. Ciarlet and M. Roseau. V, 422 pages. 1984.

Vol. 196: WOPPLOT 83. Parallel Processing: Logic, Organization and Technology. Proceedings, 1983. Edited by J. Becker and I. Eisele. V, 189 pages. 1984.

Vol. 197: Quarks and Nuclear Structure. Proceedings, 1983. Edited by K. Bleuler. VIII, 414 pages. 1984.

Vol. 198: Recent Progress in Many-Body Theories. Proceedings, 1983. Edited by H. Kümmel and M. L. Ristig. IX, 422 pages. 1984.

Vol. 199: Recent Developments in Nonequilibrium Thermodynamics. Proceedings, 1983. Edited by J. Casas-Vázquez, D. Jou and G. Lebon. XIII, 485 pages. 1984.

Vol. 200: H.D. Zeh, Die Physik der Zeitrichtung. V, 86 Seiten. 1984.

Vol. 201: Group Theoretical Methods in Physics. Proceedings, 1983. Edited by G. Denardo, G. Ghirardi and T. Weber. XXXVII, 518 pages. 1984.

Vol. 202: Asymptotic Behavior of Mass and Spacetime Geometry. Proceedings, 1983. Edited by F. J. Flaherty. VI, 213 pages. 1984.

Vol. 203: C. Marchioro, M. Pulvirenti, Vortex Methods in Two-Dimensional Fluid Dynamics. III, 137 pages. 1984.

Vol. 204: Y. Waseda, Novel Application of Anomalous (Resonance) X-Ray Scattering for Structural Characterization of Disordered Materials. VI, 183 pages. 1984.

Vol. 205: Solutions of Einstein's Equations: Techniques and Results. Proceedings, 1983. Edited by C. Hoenselaers and W. Dietz. VI, 439 pages. 1984.

Vol. 206: Static Critical Phenomena in Inhomogeneous Systems. Edited by A. Pękalski and J. Sznajd. Proceedings, 1984. VIII, 358 pages. 1984.

Vol. 207: S. W. Koch, Dynamics of First-Order Phase Transitions in Equilibrium and Nonequilibrium Systems. III, 148 pages. 1984.

Vol. 208: Supersymmetry and Supergravity/Nonperturbative QCD. Proceedings, 1984. Edited by P. Roy and V. Singh. V, 389 pages. 1984.

Vol. 209: Mathematical and Computational Methods in Nuclear Physics. Proceedings, 1983. Edited by J. S. Dehesa, J. M. G. Gomez and A. Polls. V, 276 pages. 1984.

Vol. 210: Cellular Structures in Instabilities. Proceedings, 1983. Edited by J. E. Wesfreid and S. Zaleski. VI, 389 pages. 1984.

Vol. 211: Resonances – Models and Phenomena. Proceedings, 1984. Edited by S. Albeverio, L. S. Ferreira and L. Streit. VI, 369 pages. 1984.

Vol. 212: Gravitation, Geometry and Relativistic Physics. Proceedings, 1984. Edited by Laboratoire "Gravitation et Cosmologie Relativistes", Université Pierre et Marie Curie et C.N.R.S., Institut Henri Poincaré, Paris. VI, 336 pages. 1984.

Vol. 213: Forward Electron Ejection in Ion Collisions. Proceedings, 1984. Edited by K.O. Groeneveld, W. Meckbach and I.A. Sellin. VII, 165 pages. 1984.

Vol. 214: H. Moraal, Classical, Discrete Spin Models. VII, 251 pages. 1984.

Vol. 215: Computing in Accelerator Design and Operation. Proceedings, 1983. Edited by W. Busse and R. Zelazny. XII, 574 pages. 1984.

Vol. 216: Applications of Field Theory to Statistical Mechanics. Proceedings, 1984. Edited by L. Garrido. VIII, 352 pages. 1985.

Vol. 217: Charge Density Waves in Solids. Proceedings, 1984. Edited by Gy. Hutiray and J. Sólyom. XIV, 541 pages. 1985.

Vol. 218: Ninth International Conference on Numerical Methods in Fluid Dynamics. Edited by Soubbaramayer and J.P. Boujot. X, 612 pages. 1985.

Vol. 219: Fusion Reactions Below the Coulomb Barrier. Proceedings, 1984. Edited by S.G. Steadman. VII, 351 pages. 1985.

Vol. 220: W. Dittrich, M. Reuter, Effective Lagrangians in Quantum Electrodynamics. V, 244 pages. 1985.

Vol. 221: Quark Matter '84. Proceedings, 1984. Edited by K. Kajantie. VI, 305 pages. 1985.

Vol. 222: A. García, P. Kielanowski, The Beta Decay of Hyperons. Edited by A. Bohm. VIII, 173 pages. 1985.

Vol. 223: H. Saller, Vereinheitlichte Feldtheorien der Elementarteilchen. IX, 157 Seiten. 1985.

Vol. 224: Supernovae as Distance Indicators. Proceedings, 1984. Edited by N. Bartel. VI, 226 pages. 1985.

Vol. 225: B. Müller, The Physics of the Quark-Gluon Plasma. VII, 142 pages. 1985.